Eduqas
Biology
A Level Year 1 & AS

Study and Revision Guide

Neil Roberts

Published in 2016 by Illuminate Publishing Limited, an imprint of Hodder Education, an Hachette UK Company, Carmelite House, 50 Victoria Embankment, London EC4Y 0DZ

Orders: Please visit www.illuminatepublishing.com
or email sales@illuminatepublishing.com

British Library Cataloguing in Publication Data

A catalogue record for this book is available from the British Library
ISBN 978-1-908682-64-2

Printed by Ashford Colour Press, UK

11.22

The publisher's policy is to use papers that are natural, renewable and recyclable products made from wood grown in sustainable forests. The logging and manufacturing processes are expected to conform to the environmental regulations of the country of origin.

This material has been endorsed by Eduqas and offers high quality support for the delivery of Eduqas qualifications. While this material has been through a Eduqas quality assurance process, all responsibility for the content remains with the publisher.

WJEC examination questions are reproduced by permission from WJEC.

Editor: Geoff Tuttle
Cover and text design: Nigel Harriss
Text and layout: Neil Sutton, Cambridge Design Consultants

Acknowledgments

For Isla and Lucie.
The author would like to thank the editorial team at Illuminate Publishing for their support and guidance.

About the author

Neil Roberts is a former Head of Biology and has over 20 years teaching experience in universities, schools and colleges in England and Wales, and was an experienced principal examiner for a major awarding body. In 2009, he was awarded with a fellowship of the Royal Society of Biology.

Contents

How to use this book

Knowledge and Understanding

The first section of the book covers key knowledge required for the examination. There are notes on:

- Component 1, Basic Biochemistry and Cell Organisation
- Component 2, Biodiversity and Physiology of Body Systems.

Content		Eduqas AS Biology	Eduqas A Level Biology
Basic Biochemistry and Cell Organisation	*covers*	Component 1 1.1 1.2 1.3 1.4 1.5 1.6	Core concepts 1 2 3 4 5 Component 2 2.2
Biodiversity and Physiology of Body Systems	*covers*	Component 2 2.1 2.2 2.3 2.4	Component 2 2.1 Component 3 3.1 3.2 3.3

Your practical skills will be tested in the exam papers. Examples and exam tips have been included in the guide to help you prepare.

You will also find:

- **Key terms**: many of the terms in the Eduqas specification can be used as the basis of a question, so we have highlighted those terms and offered definitions.

- **Quickfire/Extra questions**: are designed to test your knowledge and understanding of the material as you go along.

- **Pointer and Grade boost**: offer extra examination advice to improve your exam technique and raise your exam performance.

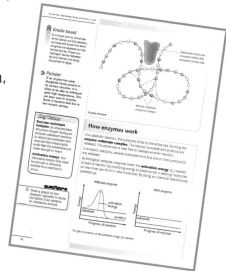

Exam Practice and Technique

The second section of the book covers the key skills for examination success and offers you examples based on suggested model answers to possible examination questions. First, you will be guided into an understanding of how the examination system works, an explanation of Assessment Objectives and how to interpret the wording of examination questions and what they mean in terms of exam answers.

This is followed by a selection of examination and specimen questions with actual student responses. These offer a guide as to the standard that is required, and the commentary will explain why the responses gained the marks that they did.

It is a good idea to split the course down into manageable chunks, complete revision notes as you go along, and have a go at as many questions as you can. The real key to success is practice, practice, practice past paper questions, so I advise that you look at www.eduqas.co.uk for sample papers and past papers. A level is such a big jump from GCSE; you really need to start working for the exams from day 1!

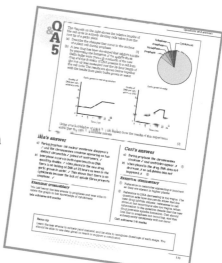

Good luck with your revision,

Dr Neil Roberts

Component **1**

Knowledge and Understanding

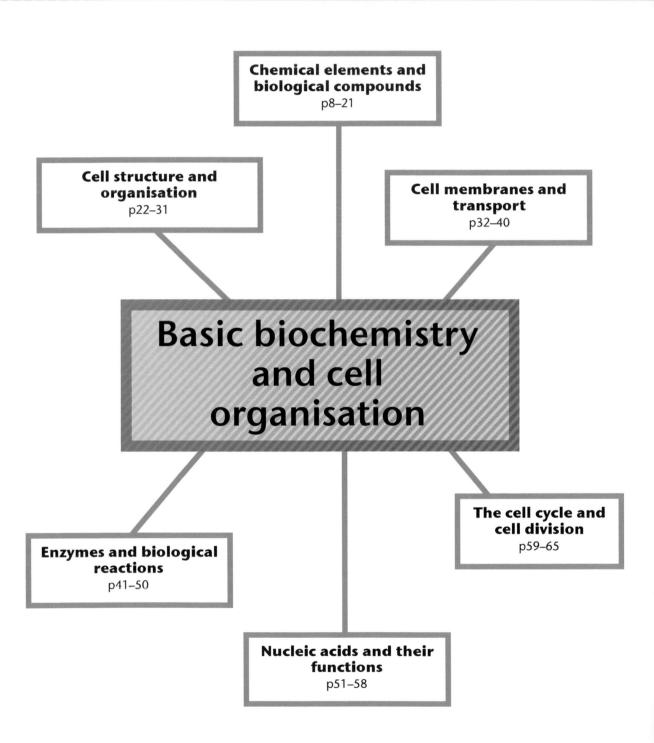

Chemical elements and biological compounds
p8–21

Cell structure and organisation
p22–31

Cell membranes and transport
p32–40

Basic biochemistry and cell organisation

Enzymes and biological reactions
p41–50

The cell cycle and cell division
p59–65

Nucleic acids and their functions
p51–58

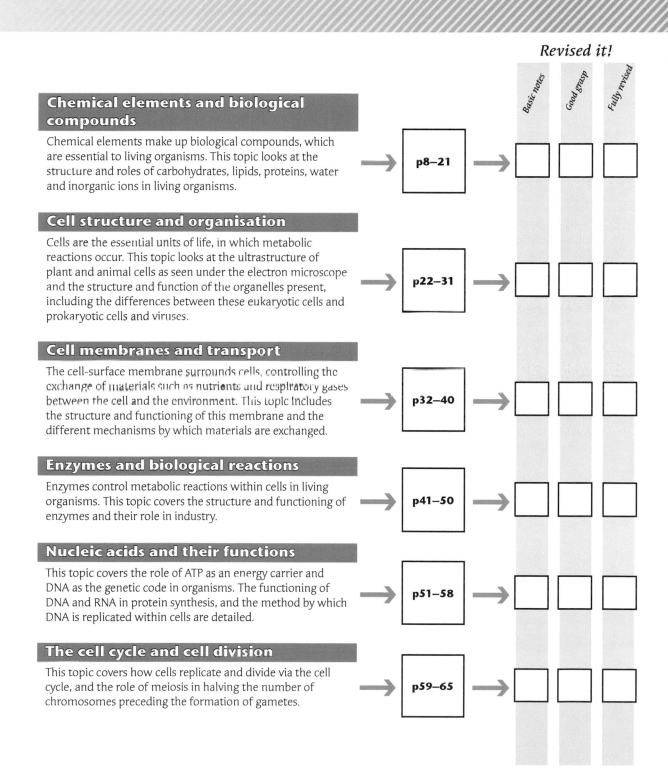

Key Terms

Condensation: the removal of a water molecule and the formation of a covalent bond between two biochemical groups, e.g. glucose + glucose = maltose + water.

Hydrolysis: the breaking down of large molecules into smaller ones by the addition of a molecule of water, e.g. lactose + water = glucose + galactose.

≫ Pointer

You should be able to recognise the structural formulae of the main biological molecules and show how bonds are formed and broken during **condensation** and **hydrolysis** reactions. You do not need to be able to reproduce them.

1.1 Chemical elements and biological compounds

Inorganic ions

A variety of inorganic ions is required for many cellular processes including muscle contraction and nervous coordination. Also known as electrolytes, some are needed in minute amounts (micronutrients), e.g. zinc, and others in small amounts (macronutrients).

Roles of inorganic ions

Inorganic ion	Role	Diagram
Magnesium (Mg^{2+})	Constituent of chlorophyll, and therefore needed for photosynthesis. When lacking, leaves appear yellow (chlorosis).	
Iron (Fe^{2+})	Constituent of haemoglobin, so is involved in transport of oxygen. A diet deficient in iron can lead to anaemia.	
Calcium (Ca^{2+})	Structural component of bones and teeth (phosphate also required).	
Phosphate (PO_4^{3-})	Needed for making nucleotides including ATP. A constituent part of phospholipids in cell membranes.	

Remember: an ion is a charged atom or molecule that has gained or lost electron(s).

Water

Water is vital to life on Earth: it makes up between 65% and 95% by mass of most organisms, allows important reactions to take place, and forms a habitat that covers over 70% of the Earth's surface.

Many of water's properties stem from its basic structure: it is a **dipolar** molecule, i.e. has a positively charged end (hydrogen) and a negatively charged end (oxygen), but has no overall charge. **Hydrogen bonds** easily form between the hydrogen on one molecule and the oxygen on another, and although individually they are weak, collectively they make it difficult to separate molecules from each other. This gives rise to many of water's properties. Water is an excellent solvent: due to its dipolar nature it attracts charged particles and other polar molecules allowing them to dissolve.

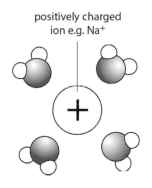

positively charged
ion e.g. Na$^+$

oxygen (δ^-) atom faces the ion

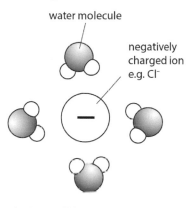

water molecule

negatively charged ion e.g. Cl$^-$

hydrogen (δ^+) atoms face the ion

Water molecules arrange themselves around ions in solution (where δ refers to a partial charge)

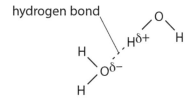

hydrogen bond

Water molecules showing hydrogen bonding

Grade boost

The majority of water's properties arise from its dipolar nature and hydrogen bonding.

quickfire

① Explain why water is able to dissolve sodium chloride.

quickfire

② Why is water said to be dipolar?

Property	Biological importance
Solvent	It is involved in many biochemical reactions, e.g. hydrolysis and condensation. Allows polar molecules, e.g. glucose, and ions, to dissolve. It acts as a transport medium, e.g. blood.
High specific heat capacity	As a large amount of heat energy is needed to increase the temperature of a body of water (due to large numbers of hydrogen bonds that need to be broken), large fluctuations in temperature are prevented. Aquatic environments are therefore relatively thermally stable.
High latent heat of vaporisation	Large amounts of heat energy are needed to vaporise water, so it is often used as a cooling mechanism, e.g. sweating in mammals.
Metabolite	It is involved in many biochemical reactions, e.g. hydrolysis and condensation, and as a reactant in photosynthesis.
Cohesion	Water molecules attract each other and form hydrogen bonds between themselves. This allows water to be drawn up the xylem vessels of trees, and creates surface tension allowing insects such as the pond skater to be supported. Water also provides support for other aquatic organisms, e.g. jellyfish.
High density	Water has a maximum density at 4°C: as a result, ice floats, and acts as an insulator preventing the water beneath from freezing completely, protecting the aquatic habitat.
Transparent	Allows light to pass through, enabling aquatic plants to photosynthesise.

quickfire

③ State the properties of water that allow:

A. insects to walk on water

B. sweat to cool an animal down and

C. water to act as a transport medium.

Grade boost

You need to be able to recognise the different types of carbohydrate and to join them together or split them apart, but you don't have to be able to draw their structure from memory.

Carbohydrates

These are small organic molecules containing carbon, oxygen and hydrogen. They act as:

- Building blocks for more complex molecules, e.g. ribose, which forms a constituent molecule of RNA.
- Sources of energy, e.g. glucose.
- Energy storage molecules, e.g. glycogen and starch.
- Structural support, e.g. cellulose and chitin.

Monosaccharides

All monosaccharides are sweet tasting and soluble in water. This group contains single sugars that all contain carbon, hydrogen and oxygen in the following proportions: $(CH_2O)_n$ where n is a number between 3 and 6. The triose sugars (n = 3) are important in respiration pathways. The pentose sugars (n = 5) such as ribose and deoxyribose are important constituents of ribonucleic acid (RNA) and deoxyribonucleic acid (DNA).

Glucose is a hexose sugar (n = 6) and is the starting material for respiration, and the building block of glycogen and other polypeptides. Other hexose sugars include galactose and fructose.

Substances that have the same formula but different structures are known as **isomers**. Glucose exists as two isomers: α-glucose and β-glucose, which differ from each other only in the positioning of the hydroxyl group on the no.1 carbon atom where α-glucose has the hydroxyl (OH) group in the down position, whereas β-glucose has the hydroxyl group in the up position. This affects the way in which they join to other molecules.

Structural formula of a triose sugar

> ### Grade boost
> Count the number of carbon atoms to determine what type of monosaccharide it is.

> ### Key Term
> **Isomer**: molecules with the same chemical formula, but with a different arrangement of atoms.

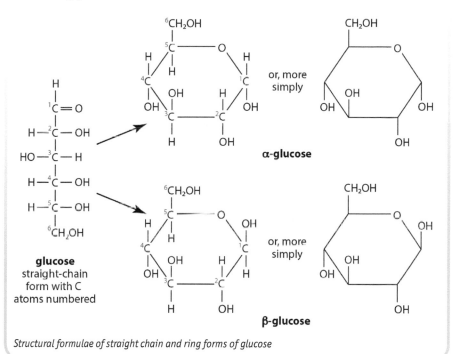

Structural formulae of straight chain and ring forms of glucose

> ### quickfire
> ④ Identify from A, B and C which one is a triose sugar, glucose, and a pentose sugar.

Disaccharides

They are formed by joining two monosaccharides together, involving the loss of a molecule of water and the formation of a glycosidic bond, via a condensation reaction.

Formation of a glycosidic bond between two glucose molecules, making maltose

Breaking down disaccharides into monosaccharides involves the chemical addition of water, known as hydrolysis.

Hydrolysis of the glycosidic bond in maltose

quickfire

⑤ If the formula of glucose is $C_6H_{12}O_6$, what is the formula of maltose (glucose + glucose)?

Disaccharide	Component monosaccharides	Biological role
maltose	glucose + glucose	in germinating seeds
sucrose	glucose + fructose	transport in phloem of flowering plants
lactose	glucose + galactose	in mammalian milk

Table summarising information about disaccharides

Grade boost

Always say that a positive test involves a colour change from *blue to red*.

Pointer

The Benedict's test is a qualitative test. It cannot give a concentration of reducing sugar present.

Pointer

Using a biosensor, a quantitative measurement can be obtained, i.e. the concentration present. Furthermore, it can detect a *specific* sugar present due to the specific enzyme used in the biosensor. This is very useful in medical monitoring of conditions such as diabetes where an accurate concentration of blood glucose is required.

Testing for presence of reducing sugars

A reducing sugar donates an electron to reduce blue copper (II) ions present in copper sulphate to red copper (I) oxide.

1. Add an equal volume of Benedict's reagent (blue) to the solution being tested and strongly heat in a boiling water bath.
2. If a reducing sugar such as glucose is present, the solution will gradually turn from blue through green, yellow and orange and finally a brick-red precipitate forms.

All monosaccharides and some disaccharides, e.g. maltose and lactose, are reducing sugars.

Some disaccharides, such as sucrose, are non-reducing sugars and a negative test will be achieved as they cannot reduce copper (II) ions in copper sulphate to copper (I) oxide. To test for a non-reducing sugar:

1. Heat with hydrochloric acid, then neutralise by adding alkali slowly until any fizzing stops.
2. Add Benedict's reagent and strongly heat as before. If the solution now turns from blue to red then a non-reducing sugar is present.

Polysaccharides

When many monosaccharides combine together, the **polymer** formed is called a polysaccharide. Polysaccharides form a number of structural molecules. They are also good energy storage molecules because they are:

- unable to diffuse out of the cell
- compact in shape so much glucose can be stored in a cell
- insoluble in water, so they do not alter the water potential, and therefore have no osmotic effect
- easily hydrolysed into their constituent monosaccharides which can be used in respiration with the exception of cellulose, which is difficult to digest due to its fibrous structure.

The polysaccharides we are concerned with are all composed solely of glucose molecules, or in the case of chitin, glucose with an acetylamine group added. The difference between them is how the glycosidic bonds are formed.

Starch

Starch is the main energy store in plants, found in starch grains which are seen in most plant cells and in chloroplasts, but are more common in seeds. Sugars made in photosynthesis are stored as starch unless they are required immediately for respiration.

Starch is made up of many α glucose molecules bonded together, and consists of two polymers, amylose and amylopectin. Amylose is linear (unbranched) with glycosidic bonds forming between the first carbon atom (C1) on molecule 1 and the fourth carbon atom (C4) on molecule 2. these are referred to as 1-4 glycosidic bonds. This is repeated forming a straight chain, which then coils into a single helix.

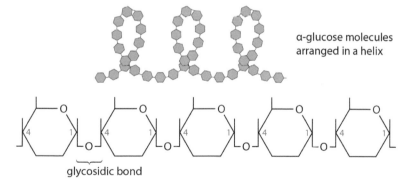

α-glucose molecules arranged in a helix

glycosidic bond

Structure of a molecule of amylose

Amylopectin is a branched molecule and fits inside the amylose. It contains 1-4 glycosidic bonds and 1-6 glycosidic bonds. When a glycosidic bond forms between the first carbon atom on one glucose molecule and the sixth carbon atom on another, this creates a side branch; 1-4 glycosidic bonds then continue on from the start of the branch. Due to its branched structure, there are more exposed ends that can be hydrolysed, which results in a more rapid release of glucose.

Grade boost

When writing about their role as storage molecules you must say storage of *energy* or *glucose*.

Key Term

Polymer: a large molecule made up of many repeating units (monomers) bonded together.

quickfire

⑥ Draw a sketch to compare amylopectin and amylose.

⟫ Pointer

The enzyme amylase breaks the glycosidic bonds in amylose and amylopectin releasing maltose.

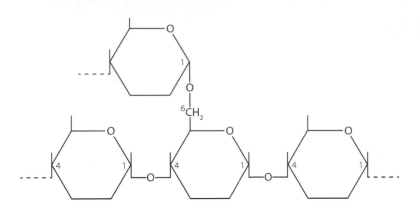

Diagram to show 1-4 and 1-6 bonds in amylopectin and glycogen

Testing for the presence of starch

Iodine solution (iodine dissolved in an aqueous solution of potassium iodide) reacts with any starch present in a sample, resulting in a colour change from orange-brown to blue-black. The depth of resulting blue-black colour seen may give an indication of concentration, but this is unreliable because as temperature increases so does the colour intensity.

Glycogen

This is the main storage product in animals and is very similar to amylopectin, differing only in that glycogen molecules are more branched than the amylopectin molecules.

Cellulose

Cellulose is a structural polysaccharide and is the most abundant organic molecule on Earth due to its presence in plant cell walls.

Cellulose consists of many β-glucose units bonded together with adjacent glucose molecules rotated by 180° forming long straight parallel chains that are cross-linked to each other by hydrogen bonds. These become tightly cross-linked to form bundles called microfibrils, which in turn are arranged into bundles called fibres. Despite their strength, the gaps between cellulose fibres in plant cell walls make them freely permeable, allowing water and solutes to penetrate through to the cell membrane.

⟰ Grade boost

Both starch and glycogen are readily hydrolysed to α-glucose, which is soluble and can then be transported to areas where energy is needed.

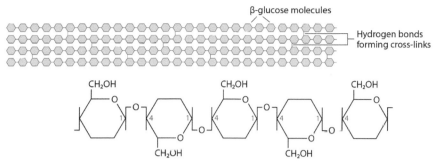

Structure of a molecule of cellulose

Grade boost

Make sure you understand that adjacent glucose molecules rotate by 180°, allowing hydrogen bonds to be formed between the hydroxyl groups of adjacent parallel chains.

Chitin

Chitin is a structural polysaccharide found in the exoskeleton of arthropods (e.g. insects) and cell walls of fungi due to its strength, lightness and waterproof properties. It has a similar structure to cellulose with many long parallel chains of β-glucose molecules (with added acetylamine group) cross-linked to each other by hydrogen bonds forming microfibrils, due to adjacent glucose molecules which are rotated by 180° in a similar way to those found in cellulose.

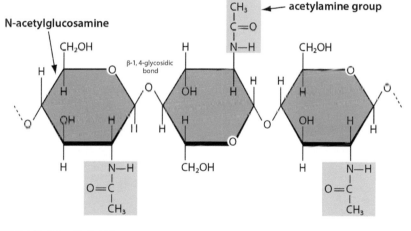

Structure of a molecule of chitin

Lipids

Lipids are non-polar compounds and so are insoluble in water. They contain:

- carbon
- hydrogen (much more than carbohydrates)
- oxygen (much less than carbohydrates).

Triglycerides are formed by the combination of one glycerol molecule and three molecules of fatty acids via a condensation reaction whereby three molecules of water are removed and an ester bond is formed between the glycerol and fatty acid.

 Pointer

An important chemical property of lipids is that they are insoluble in water but dissolve in organic solvents such as acetone (propanone) and alcohols. These can be used to dissolve the phospholipids found in cell membranes.

CH_2OH $HOOC$ — Fatty acid 1 CH_2OOC — Fatty acid 1

$CHOH$ $HOOC$ — Fatty acid 2   ⟶ $CHOOC$ — Fatty acid 2

CH_2OH $HOOC$ — Fatty acid 3 CH_2OOC — Fatty acid 3

glycerol   +   3 fatty acids   ⟶   triglyceride

Condensation reaction between glycerol and 3 fatty acids forming a triglyceride

The differences in the properties of different fats and oils come from variations in the fatty acids. If the hydrocarbon chain has no carbon–carbon double bonds then the fatty acid is described as saturated because all the carbon atoms are linked to the maximum possible number of hydrogen atoms. That is, they are saturated with hydrogen atoms. These fats are semi-solid at room temperature and are useful in the storage of fats in mammals. A high intake of fat, notably saturated fats, is a contributory factor in heart disease. Animal lipids are often saturated, whereas plant lipids are often unsaturated and occur as oils, such as olive oil and sunflower oil. Where just one carbon–carbon double bond is present they are referred to as monounsaturated, whereas with the presence of many carbon–carbon double bonds they are referred to as polyunsaturated. Where the carbon–carbon double bonds are found, the straight chain fatty acid may kink.

saturated fatty acid

unsaturated fatty acids

Structure of fatty acids

Waxes

Waxes are a type of lipid that melt at temperatures above 45°C. They perform a waterproofing role in both animals and plants, e.g. in the leaf cuticle.

Roles of lipids

Lipids play a major role in the structure of plasma membranes, and are the major component of the myelin sheath that surrounds neurones. The presence of myelin increases the speed at which nerve impulses propagate along the neurone.

Other roles of lipids include:

Property	Role
Energy reserve	Lipids make excellent energy reserves in both plants and animals. This is because they contain more carbon–hydrogen bonds than carbohydrates. One gram of fat, when oxidised, yields approximately *twice* as much energy as the same mass of carbohydrate.
Thermal insulation	When stored under the skin it acts as an insulator against heat loss.
Protection	Fat is often stored around delicate internal organs such as kidneys providing protection against physical damage.
Metabolic water	Triglycerides produce a lot of metabolic water when oxidised. This is important in desert animals such as the kangaroo rat, which survives on metabolic water from the respiration of its fat intake.
Waterproofing	Fats are insoluble in water and are important in land organisms such as insects where the waxy cuticle cuts down water loss. Water loss is then only usually possible via the stomata through the process known as transpiration.

Phospholipids

Phospholipids are a special type of lipid where one of the fatty acid tails is replaced by a phosphate group. This creates a molecule where one end is soluble in water, and the other is not. The fatty acid part is non-polar and insoluble in water so is referred to as being **hydrophobic**. The glycerol part and the phosphate group are polar and dissolve in water: this is said to be **hydrophilic**.

Phospholipids are important in the formation and functioning of plasma membranes in cells, see page 32.

quicKfire

⑧ Suggest why plants store energy as lipids.

quicKfire

⑨ What is meant by metabolic water?

Key Terms

Hydrophobic: water hating, i.e. cannot interact with water due to the lack of any charge on the molecule.

Hydrophilic: water loving, i.e. can interact with water due to the presence of charge on the molecule.

⑩ State two differences between a phospholipid and a triglyceride.

⑪ State through which part of the membrane a lipid soluble molecule would pass.

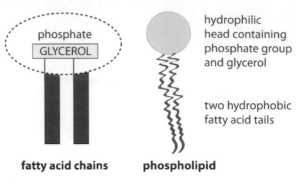

Structure of a phospholipid

Test for fats and oils

In the Lipid Emulsion Test a sample is mixed with ethanol to dissolve any lipids present (lipids are soluble in organic solvents such as ethanol). Then the sample is shaken with an equal volume of water. This causes the dissolved lipids to fall out of solution as they are insoluble in water, giving the sample a cloudy white emulsion appearance.

Implications of saturated fats on human health

Atherosclerosis is the buildup of fatty deposits or plaques called atheromas within artery walls, as a result of low-density lipoproteins (LDL) from a diet high in saturated fats. This leads to a narrowing of the arteries. As the arteries narrow, they lose their elasticity and begin to restrict blood flow, which limits oxygen delivery to the heart, which can result in angina, and eventually a heart attack. Atheromas can cause the endothelial lining to rupture, which causes a clot to form (thrombosis), which can also cause strokes.

It has been shown that diets with a higher proportion of unsaturated fats, combined with exercise, result in the body manufacturing more high-density lipoproteins (HDL) which carry harmful fats to the liver for disposal. The higher the ratio of HDL:LDL in a patient's blood, the lower the risk of cardiovascular disease.

Proteins

Proteins are large compounds built up of sub-units called amino acids. About 20 different amino acids are used to make up proteins, and it is the order of these amino acids that determines the protein's structure and hence its function.

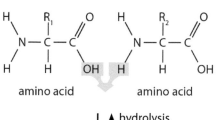

Generalised structure of an amino acid

Proteins contain:

- carbon
- hydrogen
- oxygen
- nitrogen
- sometimes sulphur.

Grade boost

You may need to identify an R group/carboxylic acid/amino group from a full structural formula of an amino acid.

quickfire

⑫ Name the bond produced in a condensation reaction between two amino acids.

Formation of a peptide bond

Proteins are built up from a linear sequence of amino acids. A condensation reaction occurs between the amino group of one amino acid and the carboxyl group of another, eliminating water. The bond that is formed is called a peptide bond and the resulting compound is a dipeptide.

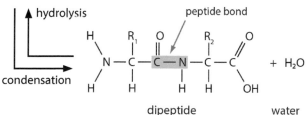

Formation of a dipeptide

Grade boost

The bond formed between two amino acids is a peptide bond NOT a dipeptide bond!

>> *Pointer*

Proteins carry out a wide range of functions. They act as enzymes, antibodies, transport proteins, hormones and structural proteins.

>> *Pointer*

Collagen is an example of a quaternary protein that does not have a tertiary structure. Three alpha helices are wound round each other to produce the structural protein.

quickfire

⑬ List the four bonds present in the tertiary structure of a protein in order of weakest to strongest.

Protein structure

Proteins are very large molecules and consist of long chains of many amino acids joined together. These chains are called polypeptides.

Protein structure	Diagram
Primary structure is the order of amino acids in a polypeptide chain. It is determined by the DNA sequence on one strand of the DNA molecule.	met — arg — lys — arg — tyr — phe — etc.
The *secondary structure* involves folding of the primary structure into a 3D shape which is held together by hydrogen bonds between =O on the –COOH group and the H on the NH_2 groups. This creates two shapes: the α helix and β pleated sheet.	hydrogen bond; - - - hydrogen bond; one amino acid; polypeptide chain; hydrogen bond
The *tertiary structure* forms from folding of the α helix into a more compact shape, which is maintained by disulphide, ionic, hydrogen bonds, and hydrophobic interactions. It gives globular proteins, e.g. enzymes, their shape.	
The *quaternary structure* arises from a combination of two or more polypeptide chains in tertiary form combined. These are often associated with non-protein groups and sometimes form large, complex molecules, e.g. haemoglobin.	α_1 polypeptide; haem (non-protein, iron-containing group); β_2 polypeptide; β_1 polypeptide; α_2 polypeptide

The roles that proteins perform depend on their molecular shape. They are classified into two groups:

- *Fibrous proteins* perform structural functions. They consist of polypeptides in parallel chains or sheets with numerous cross-linkages to form long fibres, for example keratin (in hair). Fibrous proteins are insoluble in water, strong and tough. Collagen provides tough properties needed in tendons. A single fibre consists of three polypeptide chains twisted around each other like a rope. These chains are linked by cross-bridges, making a very stable molecule.

- *Globular proteins* perform a variety of different functions – enzymes, antibodies, plasma proteins and some are hormones e.g. insulin. These proteins are compact and folded as spherical molecules. They are soluble in water. Haemoglobin consists of four folded polypeptide chains, at the centre of each of which is an iron-containing group called haem. Each protein has a unique and specific shape.

Test for protein – the Biuret test

1. Add a few drops of Biuret reagent to your sample.
2. The presence of a protein is shown by a colour change from blue to purple.

Summary of tests for basic food groups

Food	Test	Positive result
Reducing sugar	Add few drops of Benedict's reagent and heat	Turns from blue to brick red
Non-reducing sugar	Hydrolyse with HCl, neutralise, then add few drops of Benedict's reagent and heat	Turns from blue to brick red
Starch	Add few drops of iodine solution	Turns from orange-brown to blue-black
Protein	Add few drops of Biuret reagent	Turns from blue to purple
Lipid	Add ethanol and an equal volume of water and shake	Turns cloudy

Grade boost

The Biuret test works by detecting the presence of peptide bonds and is qualitative. To detect the concentration of a specific protein a biosensor would be required; however, the use of a colorimeter may give an idea of relative concentration of any non-specific proteins present in a sample.

Grade boost

Candidates often confuse the Biuret test with Benedict's and refer to heating. This will lose you the mark.

quickfire

⑭ Classify the following examples of proteins:

A. enzyme
B. collagen
C. insulin.

1.2 Cell structure and organisation

Cell structure

When studying cells it is important to have a concept of size, and what we use to measure structures within cells. The standard unit of measurement of length is the metre, m. This would not be appropriate to measure a cell as a typical cell would measure 0.003m. Instead we use a series of smaller measurements.

>> **Pointer**
1m = 1000mm
1mm = 1000µm
1µm = 1000nm

Measurement	Symbol	How many there are in 1 metre	1 of these in metres is	What it is used to measure	Example
metre	m	1	1	larger organisms	average human is 1.75m tall
millimetre	mm	1000	10^{-3}	tissues and organs	skin varies between 0.5 and 4mm thick
micrometre	µm	1,000,000	10^{-6}	cells and organelles	typical animal cell is 30µm wide
nanometre	nm	1,000,000,000	10^{-9}	molecules	DNA helix is 2.4nm wide

Key Terms

Organelle: a specialised structure found within eukaryotic cells that carries out a specific function for the cell.

Eukaryotic cells (eukaryotes): contain DNA in chromosomes in a nucleus, and possess membrane-bound organelles, e.g. plants and animals.

Resolving power: the minimum distance by which two points must be separated in order for them to be seen as two distinct points rather than a single focused image.

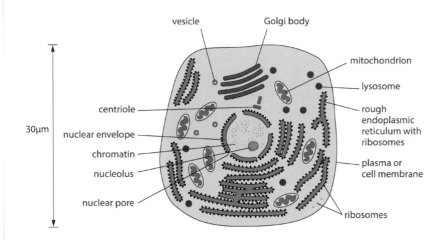

Generalised structure of an animal cell

Under the light microscope, only a few structures are easily visible within the cytoplasm. Using the electron microscope, many more structures called **organelles** are visible in **eukaryotic** cells. This is because the electron microscope uses electrons (rather than light), which have a much shorter wavelength and so the microscope has a greater magnification and higher **resolving power**. Organelles have specific roles within the cell, and are surrounded by a membrane. The membranes provide a large surface area for the transport of molecules and attachment of enzymes.

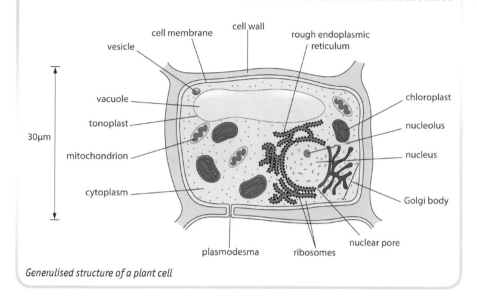

Generalised structure of a plant cell

Nucleus

The nucleus is the largest organelle present in the cytoplasm of a eukaryotic cell, which contains DNA coding for protein synthesis. The nucleoplasm contains chromatin, which condenses to form chromosomes during cell division. A double membrane, the outer membrane of which is continuous with the endoplasmic reticulum, surrounds it. The membrane has pores that allow mRNA to leave the nucleus. The nucleolus is a small spherical body found within the nucleus: it is responsible for the production of rRNA and ribosomes.

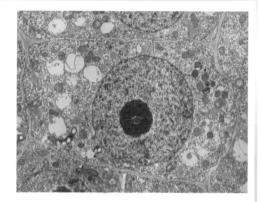

Electron micrograph of the nucleus

quickfire

① Convert the following:
   A. 2mm to micrometres
   B. 7.25 micrometres to mm
   C. 0.13 mm to micrometres.

Mitochondria

Cylindrical in shape, mitochondria are typically 1–10μm long, and are the site of aerobic respiration, producing ATP.

The inner membrane is folded into structures called cristae, which provide a large surface area for the attachment of enzymes (e.g. ATP synthetase). A fluid-filled matrix contains lipids and proteins, 70S ribosomes and a small circle of DNA. They are present in all cells, but are found in much higher number in metabolically active cells, e.g. in muscles and liver.

≫ Pointer

A cylinder has a larger surface area than a sphere of the same volume, and has a shorter distance from the edge to the centre, reducing diffusion distance and increasing respiration efficiency.

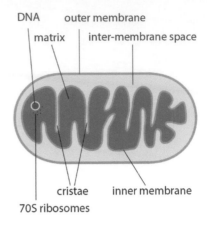

Diagram of section through a mitochondrion

Chloroplasts

Found in photosynthetic plants, chloroplasts are the site of photosynthesis. The organelle is surrounded by a double membrane, and contains a fluid-filled stroma with starch grains, 70S ribosomes and a circle of DNA. A thylakoid is a flattened membrane. Stacks of thylakoids contain the photosynthetic pigments, which include chlorophyll. Unlike the mitochondrion, the inner membrane is not folded.

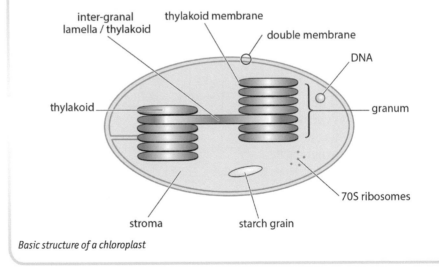

Basic structure of a chloroplast

Endoplasmic reticulum

These are a system of double membranes that form interconnected flattened fluid-filled sacs called cisternae, which are connected to the nuclear envelope. Their main role is concerned with transport of materials through the cell.

Rough endoplasmic reticulum has ribosomes attached to its outer surface, and once proteins have been synthesised at the ribosomes, they are transported via the cisternae.

Smooth endoplasmic reticulum lacks the ribosomes, and is involved with the synthesis and transport of lipids.

Ribosomes

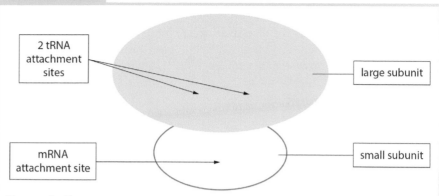

Diagram of a ribosome

Ribosomes are made from rRNA and protein, and are found within the cytoplasm. They are not surrounded by a membrane. Their role is the assembly of proteins during translation. Consisting of two subunits, the small subunit contains a mRNA attachment site, and the large subunit two tRNA attachment sites. They differ in size between cells: in eukaryotic cells ribosomes are slightly larger (80S), whilst in prokaryotic cells they are smaller at 70S.

>> *Pointer*
S = Svedburgh units.

Golgi body

Similar to endoplasmic reticulum, the Golgi is more compact in shape. It is a stack of curved cisternae. Vesicles containing polypeptides bud off the rough endoplasmic reticulum, and fuse with the Golgi. Proteins are modified and packaged into vesicles by the Golgi body for export. Golgi are also involved in the transport and storage of lipids, and the production of glycoproteins and lysosomes.

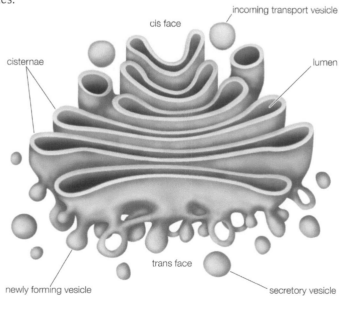

Model of a section through the Golgi body

quicKfire

② Identify the following organelle shown in the electron micrograph below:

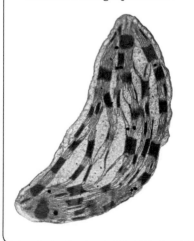

quickfire

③ Match the organelles 1–6 with their functions A–F.

1. Nucleus
2. Chloroplast
3. Ribosome
4. Golgi
5. Mitochondrion
6. Lysosome

A Site of photosynthesis
B Site of translation
C Site of transcription
D Site of aerobic respiration
E Breaks down worn out organelles
F Where proteins are modified and packaged

Lysosomes

These are small, single membrane-bound vacuoles that are pinched off from the Golgi body containing the digestive enzyme lysozyme. Their role is to digest worn out organelles within the cell, and foreign material that has been engulfed by phagocytosis, e.g. bacteria engulfed by a white blood cell.

Centrioles

Present in animal cells and protoctistans, they are noticeably absent from higher plants. Consisting of two rings of microtubules at right angles to each other, they organise the microtubules that make the spindle during cell division.

Vacuole

Within plant cells there is a large central vacuole, surrounded by the tonoplast. The main function of the vacuole is in supporting soft plant tissues, but they also store chemicals such as glucose and amino acids in the cell sap.

Cell wall

In plants the cell wall is made largely from cellulose (unlike bacteria which is peptidoglycan, and fungi which is chitin). See CELLULOSE structure on page 15.

The cell wall is important in:

- Providing strength to the cell wall, which resists the expansion of the vacuole due to osmosis, thus creating turgor and support for non-woody plants.
- Transport of water and dissolved molecules and ions through gaps in the cellulose fibres. This is known as the apoplast pathway. See page 97.
- Communication between cells via pores in the cell wall which allow strands of cytoplasm called plasmodesmata to pass. This allows water to pass via the symplast pathway. See page 97.

Differences between plant and animal cells

The table below shows the differences between plant and animal cells:

Plant cell	Animal cell
Cell wall surrounding a membrane	No cell wall, membrane only
Chloroplasts present (in cells above ground)	Chloroplasts never present
Large permanent single, central vacuole	Small, temporary vacuoles
No centrioles	Centrioles
Plasmodesmata	No plasmodesmata
Starch grains used for energy storage	Glycogen granules used for energy storage

>> *Pointer*

Many organelles work together to perform major functions within the cell, e.g. the production of proteins requires the nucleus (produces mRNA via transcription), ribosomes (produces polypeptide via translation), rough endoplasmic reticulum and Golgi body (to transport, modify and package the protein for export).

Prokaryotic cells, e.g. bacteria

These are the simplest and oldest cells on Earth, dating back over 3.5 billion years. They lack a true nucleus, instead possessing DNA loose within the cytoplasm, and have no membranous organelles, though some have infoldings of the membrane called mesosomes, where respiration is thought to occur. The cell wall is made from peptidoglycan, and the ribosomes are slightly smaller than those of eukaryotes, at 70S. Some bacteria contain plasmids (small rings of DNA) that contain antibiotic resistance genes

Grade boost

Be prepared to compare a prokaryotic cell with an eukaryotic cell.

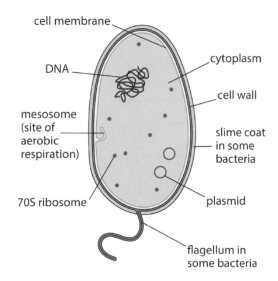

Diagram of a generalised bacterial cell

quickfire

④ Complete the table to compare prokaryotes and eukaryotes.

Prokaryotic cell	Eukaryotic cell
e.g. bacteria and blue-green algae	e.g. plants, animals, fungi and protoctists
No membrane-bound organelles	
	Ribosomes are larger (80S) lie free and attached to membranes, e.g. rough ER
	DNA located on chromosomes within the nucleus
No nuclear membrane	
	Cell wall in plants made of cellulose. In fungi it is made of chitin

Viruses

They do not possess a cytoplasm, organelles or any chromosomes: just a core of nucleic acid (which can be DNA or RNA) surrounded by a protein coat, called the capsid. This inert 'virion' is incapable of reproducing or synthesising proteins without the use of a host's cytoplasm. It is when they burst out of cells and reinfect healthy cells that damage occurs. Examples include the tobacco mosaic virus that causes tobacco mosaic disease, and HIV that causes HIV-AIDS.

Grade boost

Remember to always measure in mm to avoid errors when converting between the different units of measurements. You need to be able to convert mm to µm.

Key Term

Magnification: how many times bigger the image is, compared to the object.

Examining cells

Measuring from electron micrographs

To do this, you will need a formula:

Magnification = $\dfrac{\text{size of image}}{\text{size of object}}$

This can be easily rearranged to calculate the size of the object when given the **magnification** and a photograph to measure

Size of object = $\dfrac{\text{size of image}}{\text{size of magnification}}$

Worked example

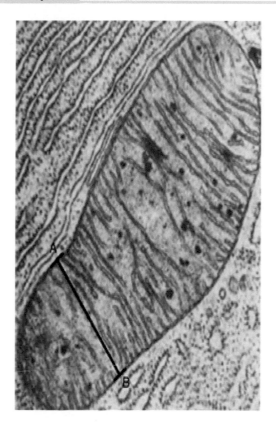

Electron micrograph of mitochondrion to measure

The width of the mitochondrion shown above was measured at its narrowest point A–B and found to be 1.0μm. Calculate the magnification of the electron micrograph.

Magnification = $\dfrac{\text{size of image}}{\text{size of object}}$

Measuring the image A–B is 35mm

This needs to be multiplied by 1000 to convert to μm so both sizes being compared are in μm

$$= \frac{35 \times 1000}{1.0}$$

$$= 35{,}000 \times \text{magnification.}$$

Grade boost

You will need to be able to identify electron micrographs of different organelles and from these calculate either the magnification of the image or the size of object in real life.

Grade boost

When comparing size of image with size of object, both sizes should be in the same units – so be prepared to convert! A good rule to remember is measure in mm then ×1000 to convert to μm.

Grade boost

Always show your working as credit can sometimes be given even if your answer is incorrect.

Grade boost

You will not be asked to describe nervous tissue in the exams.

Levels of organisation

Single-celled organisms carry out all life processes within the one cell, whereas multicellular organisms need to possess specialised cells that form **tissues** and **organs** to do this. Stem cells are undifferentiated (non-specialised) cells in the embryo that can differentiate to form any tissue. In mammals there are four main types of tissue: nervous, connective, muscle and epithelial.

Tissues

1. Connective tissue – supports, connects or separates the different types of tissues and organs of the body. Cells are contained within an extracellular fluid or matrix and may be surrounded by elastic or collagenous fibres, e.g. tendons and blood.

2. Muscle tissue – three types:
   - Skeletal muscle has bands of long cells or fibres giving powerful contraction and is used for locomotion in mammals.
   - Smooth muscle has individual spindle-shaped cells which contract rhythmically but are not powerful so they are found in the walls of blood vessels, the digestive and respiratory tracts.
   - Cardiac muscle cells have stripes but lack long fibres. They contract rhythmically and with some force but do not tire easily.

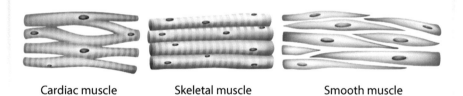

Cardiac muscle   Skeletal muscle   Smooth muscle

Diagram to show three types of muscle fibre

3. Epithelial tissue – covers and lines the body, e.g. lining the intestines and trachea and covers our body as part of our skin. All epithelial cells sit on a basement membrane, but cells do vary in shape and complexity. Examples:
   - The simplest form is simple cuboidal where cells are a cube shape and the tissue is just one cell thick. This tissue is commonly found lining kidney tubules and the ducts of glands.
   - Columnar epithelium has cells that are more rectangular and may have cilia present, e.g. lining the trachea.
   - Squamous epithelium consists of flattened cells. They are found in the alveoli and lining arteries.

Diagram of cuboidal epithelium

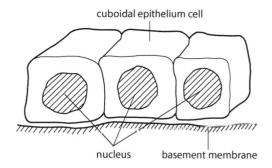

cuboidal epithelium cell

nucleus   basement membrane

Diagram of ciliated columnar epithelium

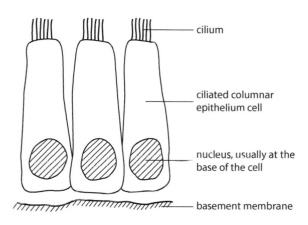

cilium

ciliated columnar epithelium cell

nucleus, usually at the base of the cell

basement membrane

Diagram of squamous epithelium

squamous epithelium cell

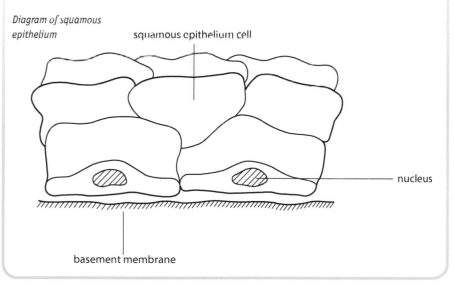

nucleus

basement membrane

quickfire

⑤ Match the tissues 1–6 with where you would find them in the body A–F.
1. Cuboidal epithelium
2. Columnar epithelium
3. Squamous epithelium
4. Smooth muscle
5. Cardiac muscle
6. Connective tissue

A. In tendons
B. Muscle of the heart
C. In the walls of blood vessels
D. Kidney tubules
E. Lining arteries
F. Lining trachea

Organs and organ systems

Cells → Tissues → Organs → Organ systems

Groups of organs working together with a particular role are called organ systems. Examples include the digestive system (stomach, ileum, colon), and the circulatory system (heart, arteries, capillaries and veins).

1.3 Cell membranes and transport

Phospholipids (see lipids on page 17)

Due to the hydrophilic polar head and hydrophobic fatty acid tails, phospholipids form a bi-layer in water: polar heads face outwards interacting with the water outside the cell and inwards interacting with the water in the cytoplasm.

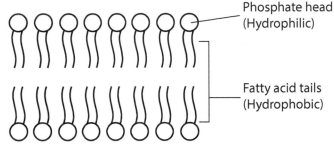

Phospholipid bilayer

In the cell membrane, phospholipids arrange themselves into this bilayer, with proteins scattered throughout. Some proteins are **extrinsic** and found on the surface of the bilayer, acting as receptors for hormones and as recognition sites, whilst others are **intrinsic** and extend across both layers, acting as channels and carrier proteins for the transport of molecules. Some proteins are enzymes, e.g. ATP synthetase in the cristae or digestive enzymes in the epithelium of the villi.

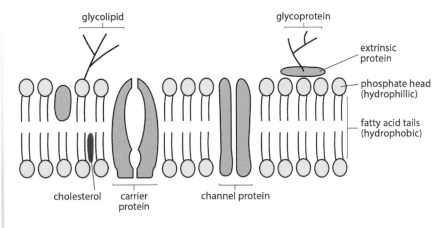

Fluid mosaic model of plasma membrane

The fluid mosaic model of the membrane structure was proposed by Singer and Nicolson in 1972. The membrane in animal cells also contains cholesterol, which stabilises it. Glycoproteins act as antigens, meanwhile glycolipids act as receptor sites for molecules such as hormones.

Key Terms

Extrinsic proteins: these are found on either surface of the membrane.

Intrinsic proteins: these are found within the membrane.

 Pointer

Fluid mosaic: fluid because phospholipids are free to move, and mosaic due to the random assortment of protein molecules.

Grade boost

Make sure you write about the structure of the membrane when you are trying to explain how different materials cross the membrane.

Transport across membranes

The properties of molecules passing across the membrane will directly affect *how* they cross it, e.g. non-polar molecules such as vitamin A and small molecules such as oxygen can dissolve in the fatty acid tails and diffuse across the membrane. Polar molecules, e.g. glucose, have to pass via a transport protein, because they cannot dissolve in the fatty acid tails.

>> *Pointer*
Passive transport does not require energy from ATP.

Diffusion

Simple diffusion is an example of passive transport whereby molecules move from a high concentration to a low concentration until they are equally distributed. Molecules are constantly moving due to their kinetic energy: any factor that increases this energy, or decreases the distance that they have to diffuse, will increase the rate of diffusion. The rate of diffusion is affected by:

- The concentration gradient: the greater the difference in the concentration of molecules in two areas, the more molecules that can diffuse in a given time so collisions with the membrane are more likely.

- Diffusion distance: it takes less time for the molecules to diffuse a shorter distance.

- The surface area of the membrane: the larger the area, the more molecules that can diffuse in a given time.

- The thickness of the exchange surface: it takes less time for the molecules to diffuse a shorter distance.

- An increase in temperature results in molecules possessing more kinetic energy so they move faster and collide with the membrane more frequently.

Diffusion is proportional to: $\dfrac{\text{surface area} \times \text{difference in concentration}}{\text{length of the diffusion path}}$

quickfire

① The graph shows how oxygen uptake in the roots of plants is affected by oxygen concentration.

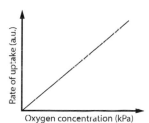

Rate of oxygen uptake in roots

Can you explain why this graph shows diffusion?

Facilitated diffusion

Polar molecules such as glucose cannot pass through the cell membrane because they are relatively insoluble in lipids. Facilitated diffusion is like simple diffusion in that it is a passive process so requires no ATP, relying instead upon the kinetic energy of the molecules involved. However, it relies on transport proteins found within the membrane that assist the movement of polar molecules across the membrane. It therefore is affected by the same factors as diffusion with one addition: ultimately the rate will be determined by the number and availability of the transport proteins involved.

There are two types of transport proteins:

1. *Channel proteins* consist of pores with a hydrophilic lining allowing charged ions and polar molecules to pass through. They are specific, and can be opened or closed to regulate the movement of particular molecules.

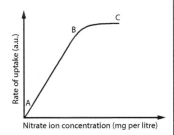
2. *Carrier proteins* allow diffusion across the membrane of larger polar molecules such as sugars and amino acids.

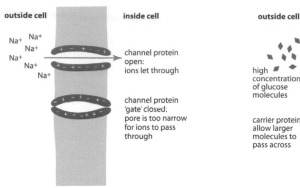

Carrier and channel proteins

Active transport

This form of transport requires energy in the form of ATP: it is therefore able to transport molecules AGAINST the concentration gradient. Anything that affects the respiratory process will affect active transport, e.g. cyanide is a respiratory inhibitor, which will prevent aerobic respiration and the production of ATP. In the absence of ATP, active transport cannot occur. Active transport also utilises carrier proteins that span the membrane, and therefore the maximum rate will be ultimately limited by the number and availability of these transport proteins.

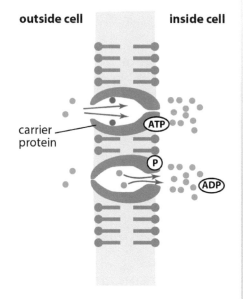

Carrier proteins change shape when transporting a molecule across a membrane

Active transport and respiratory inhibitors

The graph shows that a maximum rate of transport can still be reached when the carrier proteins are saturated. The rate of uptake is reduced with the addition of a respiratory inhibitor: active transport must be taking place as the process requires ATP.

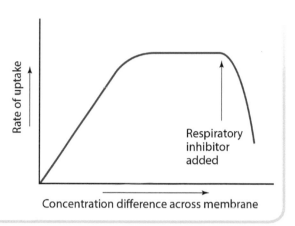

Inhibition of active transport

Rate of uptake

Respiratory inhibitor added

Concentration difference across membrane

>> *Pointer*

ATP is important in the transfer of energy and is produced during respiration. You will study this molecule in more detail in year 2 of the course.

Co-transport

Co-transport involves transporting two different molecules together, e.g. glucose and sodium ions, and is the mechanism by which glucose is absorbed in the ileum of mammals.

1. Sodium ions are actively transported out of epithelial cells lining the ileum into the blood, creating a low concentration of sodium ions within the cells.

2. The higher concentration of sodium ions in the lumen of the gut, compared to the epithelial cells, causes sodium ions to diffuse into the epithelial cells via a co-transport protein. As they do so they couple with glucose molecules carrying them with them.

3. Finally glucose molecules pass via facilitated diffusion into blood capillaries and sodium ions by active transport.

>> *Pointer*

It is the sodium ion concentration gradient that powers the movement of glucose in co-transport not ATP directly.

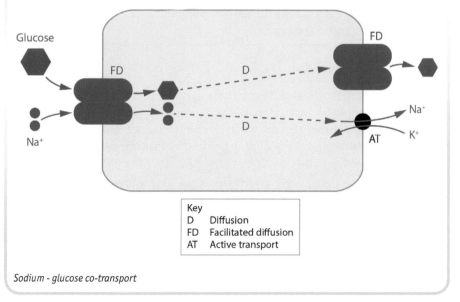

Glucose

FD

D

FD

Na^+

D

Na^+

AT

K^+

Key
D   Diffusion
FD   Facilitated diffusion
AT   Active transport

Sodium - glucose co-transport

Key Terms

Solute: any substance that is dissolved in a solvent.

Water potential (ψ): represents the tendency for water to move into or out of a system, and is the pressure created by water molecules.

Osmosis: the net passive diffusion of water molecules across a selectively permeable membrane from a region of higher water potential to a region of lower water potential.

Pressure potential (ψ_p): represents the pressure exerted by the cell contents on the cell wall. It can be 0kPa or higher.

Turgid: means that the plant cell can hold no more water, as the cell wall cannot expand further.

A **hypertonic solution**: has a lower water potential (ψ) relative to the solution inside the cell, due to the presence of solutes.

A **hypotonic solution**: has a higher water potential (ψ) relative to the solution inside the cell, due to the absence of solutes.

An **isotonic solution**: has the SAME water potential (ψ) relative to the solution inside the cell.

quickfire

④ The water potential of cell A = −200kPa, cell B= −400kPa, cell C = −300kPa. Which directions will water move by osmosis? Which movement will occur the quickest and why?

Osmosis

Most cell membranes are permeable to water and certain **solutes** only.

When there is a high concentration of water molecules they can move about freely. When a solute such as glucose is dissolved, there are proportionally fewer water molecules, so they can move about less freely. In other words, adding a solute lowers the potential for water molecules to move: it lowers the **water potential** making it more negative.

Examples of water potentials:

- pure water is 0kPa
- typical cell is −200kPa
- strong glucose solution is −1000kPa.

Osmosis and plant cells

When water enters a cell by **osmosis**, the vacuole expands, pushing the cytoplasm against the cellulose cell wall. The cell wall can only expand a little, so this creates a resistance to more water entering a plant cell by osmosis, which is known as the **pressure potential**. The cell is said to be **turgid**.

This is represented by:

$$\psi = \psi_s + \psi_p$$

water potential = solute potential + pressure potential

If a cell is placed into a solution that is **hypotonic** to the cell: water flows into the cell, because the solution has a higher water potential. If a cell is placed into a solution that is **hypertonic** to the cell: water flows out of the cell, because the solution has a lower water potential than the cell. If the cell has the same water potential (ψ) as the surrounding solution, the external solution and internal solutions are **isotonic** and there will be no <u>net</u> water movement in or out of the cell.

Turgor and plasmolysis

When a plant cell is placed in a hypertonic solution it loses water by osmosis. The vacuole shrinks and the cytoplasm will draw away from the cell wall. This process is called plasmolysis and, when complete, the cell is said to be flaccid. The point at which the cell membrane just begins to move away from the cell wall is said to be the point of incipient plasmolysis.

The water potential is now equal to the pressure and **solute potentials**, i.e.

$$\psi = \psi_p + \psi_s.$$
$$\psi_p = -\psi_s$$
$$\psi = 0$$

When the cell cannot take in any more water, it is turgid. Turgor is important to plants, especially young seedlings, because it provides support, maintains their shape and holds them upright.

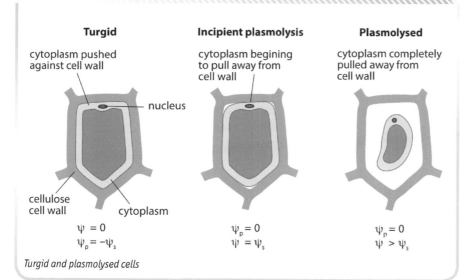

Turgid

cytoplasm pushed against cell wall

nucleus

cellulose cell wall

cytoplasm

$\psi = 0$
$\psi_p = -\psi_s$

Incipient plasmolysis

cytoplasm begining to pull away from cell wall

$\psi_p = 0$
$\psi = \psi_s$

Plasmolysed

cytoplasm completely pulled away from cell wall

$\psi_p = 0$
$\psi > \psi_s$

Turgid and plasmolysed cells

Grade boost

When writing about the movement of water, always say it is by osmosis.

Grade boost

Water potential is always negative except for the highest water potential, which is the water potential of pure water, 0kPA.

Measuring incipient plasmolysis

This can be estimated by placing plant cells into solutions with varying solute potentials, and then looking at the cells under a microscope. The percentage of plasmolysed cells is calculated using the formula below and a graph can be drawn.

$$\% \text{ plasmolysis} = \frac{\text{number of plasmolysed cells}}{\text{total number of cells observed}} \times 100$$

When the percentage plasmolysis is equal to 50%, incipient plasmolysis has been reached, and the external sucrose concentration must be equal to the internal solute concentration of the onion tissue (as there is no net movement of water overall). The solute potential can be calculated using a conversion table. In the example shown below it is -680 kPa.

Key Term

Solute potential (ψ_s): represents the osmotic strength of a solution. It is the reduction in water potential due to the presence of solute molecules. It is 0kPa or negative.

Molarity of sucrose solution (M)	Solute potential kPa
0.05	−130
0.10	−260
0.15	−410
0.20	−540
0.25	−680
0.30	−860
0.35	−970
0.40	−1120
0.45	−1280
0.50	−1450
0.55	−1620
0.60	−1800

Conversion table

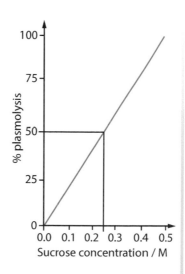

Incipient plasmolysis graph

quickfire

⑤ Water enters root hair cells by osmosis. Calculate the solute potential (Ψ_S) of the root hair cell, when there is no net movement of water, the water potential of the soil water is –100kPa and the pressure potential (Ψ_P) inside the root hair cell is +200kPa. Use the formula $\Psi = \Psi_S + \Psi_P$. Show your working and units.

Grade boost

Always remember your units and show your working with any calculation.

Grade boost

You can expect to see 10% of questions in exam papers being mathematical.

Osmosis and animal cells

Animal cells lack a cell wall, and therefore they cannot sustain a pressure potential. Because no pressure potential can exist (as animal cells will burst) the water potential is the same as the solute potential or $\Psi = \Psi_S$.

When red blood cells are placed into a hypotonic solution, water enters by osmosis and they burst: this is called haemolysis. If red blood cells are placed into a hypertonic solution, water passes out of the cells and they are said to be crenated.

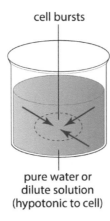

cell bursts

pure water or dilute solution (hypotonic to cell)

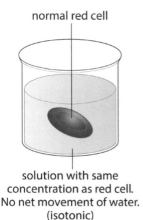

normal red cell

solution with same concentration as red cell. No net movement of water. (isotonic)

cell shrinks

concentrated solution (hypertonic to cell)

Osmosis in animal cells

Permeability of membranes

The permeability of cell membranes can be investigated by using beetroot, which has vacuoles containing a red pigment called betacyanin. The rate that betacyanin diffuses out of the vacuole through its membrane is affected by a number of factors including temperature and the presence of organic solvents.

Beetroot discs of equal size and volume were cut using a borer, then washed and blotted dry. Discs were then placed into water, and the quantity of betacyanin that leaked out through the membrane was measured using a colorimeter. The experiment was repeated at different temperatures.

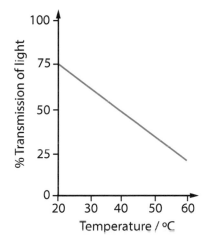

Effect of temperature on permeability of beetroot membranes

 Pointer

Colorimeters can measure the % transmission of a specific wavelength of light, e.g. 450nm, through a sample of liquid. The darker the sample is, the lower the % transmission.

 Grade boost

You need to be able to draw and explain valid conclusions using your biological knowledge from experimental data.

quickfire

⑥ What can you conclude from the results?

Beetroot vacuoles contain a red pigment called betacyanin. When beetroot discs are cut with a borer and immersed in a solution of 70% ethanol (an organic solvent) at 15°C, the red pigment begins to leak out of the cells into the ethanol turning it red.

(i) Using your knowledge of the structure of cell membranes, explain why this leakage of pigment occurs.

(ii) When the experiment was repeated at 30°C, the time taken for the ethanol to turn red decreased. Explain why.

Bulk transport

This is where the cell transports materials in bulk into the cell (endocytosis) or out of the cell (exocytosis). Endocytosis involves the engulfing of the material by infolding of the plasma membrane bringing it into the cell enclosed within a vesicle. There are two types of endocytosis:

i) *Phagocytosis* is the process by which the cell can obtain solid materials that are too large to be taken in by other methods e.g. phagocytes (white blood cells) destroy bacteria and remove cell debris by phagocytosis, and this is how amoeba feed.

ii) *Pinocytosis* is the process by which the cell can obtain liquid materials. It is similar to phagocytosis but the vesicles produced are smaller.

Exocytosis refers to substances leaving the cell after being transported through the cytoplasm in a vesicle. Digestive enzymes are often secreted in this way.

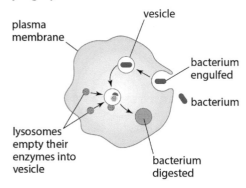

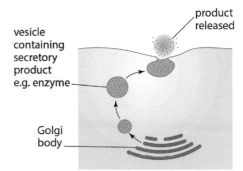

Bulk transport

1.4 Enzymes and biological reactions

Metabolism

Metabolism is a series of enzyme-controlled reactions in the body. There are TWO main types of reaction: building up reactions also called anabolic reactions or anabolism and breaking down reactions known as catabolic reactions or catabolism.

Anabolic reaction = protein synthesis where amino acids are built up into more complex polypeptides.

Catabolic reaction = digestion of proteins, where complex polypeptides are broken down into simple amino acids.

Key points about enzymes:

- They are proteins that speed up chemical reactions.
- They lower the activation energy needed for the reaction to take place.
- They don't actually take part in the reaction.
- They are only needed in small quantities.
- They can be used over and over again.
- They convert *substrates* into *products*.
- Therefore they can be described as biological catalysts.

Enzyme structure

Enzymes are complex folded polypeptide chains that are held together in a complex 3D shape. Their most basic structure, called the **primary structure**, is formed from the order of the different amino acids that are organised into chains called polypeptides. Each amino acid is joined to the next one by a **condensation** reaction, which forms a peptide bond (see page 19). This structure is then folded into either an alpha helix or beta pleated sheet, held together by hydrogen bonds called the **secondary structure**. Enzymes have a **tertiary structure** whereby the secondary structure is folded again to form a 3D shape that is held together by hydrogen, ionic, and disulphide bonds. This is important in enzymes as it creates an 'active site' where substrates can bind. The bonds holding the tertiary structure in place are susceptible to changes in temperature, pH and the action of reducing agents. Enzymes act in an aqueous environment because they are soluble and catalyse many reactions including **hydrolysis**, which we looked at earlier, see page 12.

quicKfire

① Write an equation to represent metabolism using the words anabolism and catabolism.

Key Terms

Primary structure: formed from the order of amino acids.

Condensation: reaction occurs joining two molecules together into a larger one with the elimination of water.

Secondary structure: formed from the folding of the primary structure into two main forms: the alpha helix or beta pleated sheet.

Tertiary structure: formed from the folding of the secondary structure into a 3D shape.

Hydrolysis: the breaking down of large molecules into smaller ones by the addition of a molecule of water.

Grade boost

It is important to remember which bonds are the weakest, as these will break first when enzymes are exposed to high temperatures. These are: hydrogen bonds followed by ionic bonds and lastly disulphide bridges.

Pointer

If an enzyme has many disulphide bonds present in its tertiary structure, it is likely to be able to withstand quite high temperatures. This can been seen in enzymes found in bacteria that live in hot volcanic springs.

Key Terms

Enzyme–substrate complex: an intermediate structure formed during an enzyme-catalysed reaction in which the substrate and enzyme bind temporarily, such that the substrates are close enough to react.

Activation energy: the minimum energy that must be put into a chemical system for a reaction to occur.

quickfire

② Draw a graph on the diagram opposite to show the effect of an enzyme on activation energy.

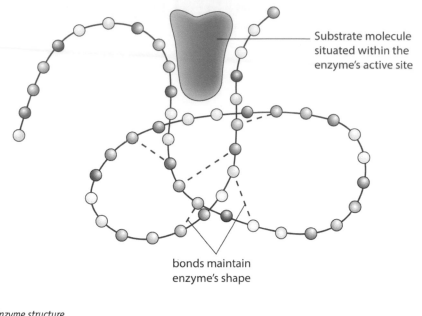

Enzyme structure

How enzymes work

In a catabolic reaction, the substrate binds to the active site, forming the **enzyme–substrate complex**. The reaction proceeds and products are released. The active site is now free to catalyse another reaction.

In anabolic reactions, several substrates bind and one or more product(s) are released.

As biological catalysts, enzymes lower the **activation energy** (E_A) needed to start a reaction by providing energy to break bonds in existing molecules so new ones can form in new molecules. By doing so, chemical reactions are speeded up.

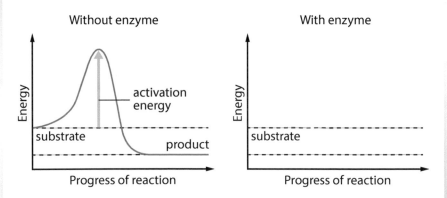

The effect of enzymes on the activation energy of a reaction

Enzymes may act intracellularly (within a cell), e.g. during protein synthesis where the formation of a peptide bond between two amino acids is catalysed, or extracellularly (outside a cell), e.g. when pancreatic amylase is released from pancreatic cells and travels to the small intestine via the pancreatic duct where it then catalyses the breakdown of starch to maltose.

Models of enzyme action

Two models have been proposed to explain how enzymes work on a molecular level. Each of them explains a different property of enzymes.

1. *Lock and key model* – the substrate has a complementary shape to the enzyme's active site, like a key fitting into a lock. This explains the specificity of many enzymes, i.e. that many only catalyse *one* substrate.

Lock and key model

2. *Induced fit model* – many observations show that in fact the enzyme's active site was being altered by the binding substrate molecule. The induced fit theory suggests that the active site is able to change slightly to accommodate the substrate a bit like a latex glove stretching to accommodate a hand. This change places strain on the substrate molecule, helping to break bonds and so lowering the activation energy. This explains why in some cases several molecules with similar shapes are able to bind to the active site. This is shown by the enzyme lysozyme, which is an anti-bacterial enzyme found in human tears and saliva. The active site consists of a groove, which closes over the polysaccharides found in the bacterial cell walls, and the enzyme molecule changes shape, which allows hydrolysis to occur.

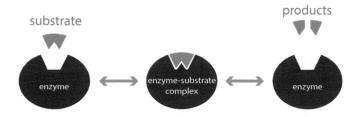

Induced fit model

Grade boost

The substrate has a complementary shape to the active site, *not* the same shape.

Factors affecting the rate of enzyme action

The rate of reaction can be considered as the number of reactions that occur per second or unit time.

Enzyme action is affected by *five* things:

1. Substrate concentration
2. Temperature
3. pH
4. Enzyme concentration
5. Presence of inhibitors.

Substrate concentration

When the substrate concentration increases in an enzyme-controlled reaction, there is a greater chance of a successful collision between the substrate and the enzyme, resulting in more enzyme–substrate complexes forming, which increases the rate of reaction. When all the enzyme active sites are occupied, a plateau is reached which represents the maximum rate of reaction for the conditions.

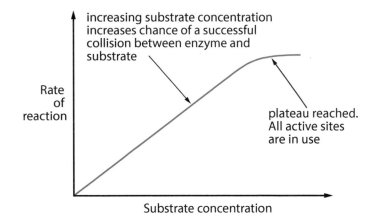

Effect of substrate concentration on the rate of reaction

Temperature

When the temperature of an enzyme and substrate increases in an enzyme-controlled reaction, both the enzyme and substrate molecules gain more **kinetic energy** and so move faster, increasing the chance of a successful collision between them. As more enzyme–substrate complexes are formed, the rate of reaction increases up to an optimum. Above this, the rate of reaction decreases rapidly as hydrogen bonds in the tertiary structure break due to increased vibrations resulting in a change to the shape of the active site – this is called **denaturing**.

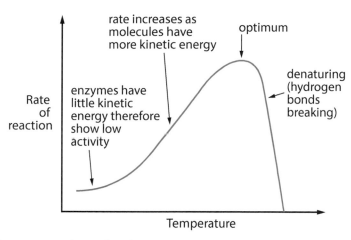

Effect of temperature on the rate of reaction

As a general rule, the rate of any enzyme-controlled reaction doubles per 10°C rise in temperature, until the optimum is reached. Therefore, the rate will have doubled at 32°C compared to 22°C. This is shown as $Q_{10} = 2$, i.e. ×2 per 10°C rise.

pH

When the pH of an enzyme increases or decreases either side of the optimum, the rate of reaction decreases. The charges on the amino acid side chains (R groups) that make up the enzyme's active site are influenced by free hydrogen (H⁺) and hydroxyl (OH⁻) ions. If too many H⁺ or OH⁻ ions are present, the substrate can be repelled from the active site, preventing it from binding. If these changes are relatively minor, then this is reversible. More excessive changes in pH will result in the ionic bonds in the tertiary structure breaking which causes denaturing by creating a permanent change to the shape of the active site.

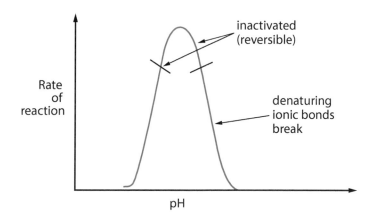

Effect of pH on the rate of reaction

Grade boost

Only describe the enzyme as being denatured when the rate of reaction is zero. The enzyme is denaturing up to this point.

Grade boost

High temperatures denature enzymes they are NOT killed!

charges on active site match those of substrate so an enzyme–substrate complex forms

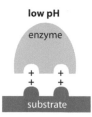
charges on active site repel substrate

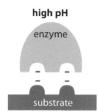
charges on active site repel substrate

Effects of PH

Use of buffers in enzyme experiments

>> *Pointer*

A buffer:
- Is a chemical that resists changes in pH.
- Neutralises excess acids or alkalis.
- Can be used to maintain the optimum pH for a given reaction.

The pH graph on page 45 shows us that the rate of an enzyme-controlled reaction is greatly influenced by small changes in pH. It is therefore essential, when carrying out any enzyme experiment (where pH is not the independent variable), that pH is controlled, ideally at its optimum, so it is not limiting the rate of reaction. This can be achieved by using a pH buffer. A buffer is a solution that can resist changes in pH by neutralising acid/alkalis that are added to the solution. In the body, we buffer the pH of the blood around 7.4 by using two chemicals – carbonic acid and bicarbonate. We look at this in more detail on page 93.

Enzyme concentration

When the substrate concentration increases in an enzyme-controlled reaction, there is a greater chance of a successful collision between the substrate and enzyme so more enzyme–substrate complexes are formed, thus increasing the rate of reaction. As long as substrate is present in excess this will continue to rise so long as there are no limiting factors.

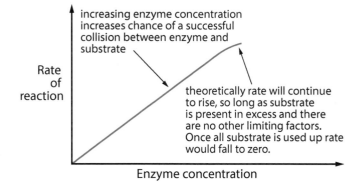

Effect of enzyme concentration on the rate of reaction

Product formation

Product formation is different from the rate of reaction as it shows the *total* product made. Once a plateau is reached, no more product is formed and the reaction has stopped. With a rate of reaction graph, the rate would drop to zero at this point.

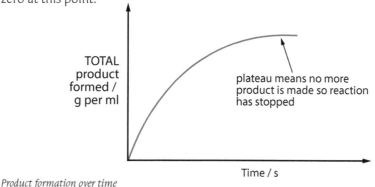

Product formation over time

Grade boost

You could be asked to use the linear relationship $y = mx + c$ on a graph.

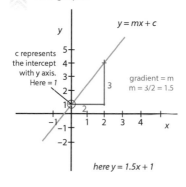

here $y = 1.5x + 1$

Inhibitors

Enzymes can be inhibited by other substances, which can either combine with the active site directly or bind to another part of the enzyme to prevent the formation of an enzyme–substrate complex. Two forms of inhibition exist, competitive and non-competitive inhibition, which may be either reversible (where inhibitor binds temporarily) or irreversible (where the inhibitor binds permanently).

quickfire

③ Plot the line on a graph where $y = -x + 2$ (*hint* $m = -1$, $c = 2$).

➤➤ Pointer

You can calculate the rate at a specific point on a graph by drawing a line at a tangent to the curve, and then calculating the gradient of that line.

Competitive inhibition

This is where a molecule has a similar shape to the substrate and so it also has a complementary shape to the active site. The first molecule to collide successfully with the active site will form a complex. By increasing the concentration of substrate, the inhibition is overcome, so long as the inhibition is reversible, as it is more likely that a substrate molecule will form an enzyme–substrate complex.

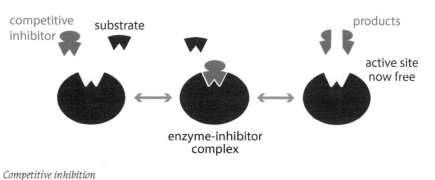

Competitive inhibition

Grade boost

In reversible competitive inhibition, the substrate and competitive inhibitor both 'compete' for the active site. It can be overcome by increasing substrate concentration.

quickⲫire

④ Describe how non-competitive and competitive inhibitors differ in the way that they attach to the enzyme.

Grade boost

In non-competitive inhibition, the inhibitor binds to an allosteric site deforming the shape of the enzyme's active site. It cannot therefore be overcome by increasing substrate concentration.

An example is the enzyme succinic dehydrogenase, which is involved in cellular respiration. It catalyses the breakdown of succinate to fumarate. Malonate is a competitive inhibitor of this enzyme due to its similar shape to the substrate.

Succinate and malonate

Non-competitive inhibition

Here, an inhibitor binds to another site on the enzyme (the allosteric site). This binding changes the shape of the active site, preventing substrate molecules from forming an enzyme–substrate complex. An example is cyanide that binds to cytochrome oxidase inhibiting respiration.

Non-competitive inhibition

End-product inhibition

This is often seen in complex metabolic pathways where several enzymes are involved. It is an example of competitive inhibition at work in cells, and prevents the build-up of the end product in the pathway, which could become harmful. In essence, the product of one reaction acts as the substrate for the next, and the end product acts as a competitive inhibitor for an enzyme earlier in the pathway. In the example shown, the end product inhibits enzyme 1: as the end product is used up in the cell, the concentration of end product falls and the concentration of the initial substrate rises, so overcoming the inhibitor's effect.

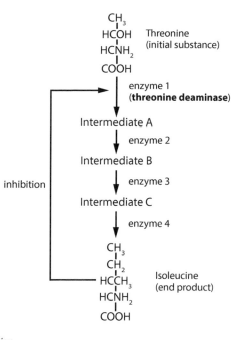

End-product inhibition

Grade boost
You should be able to list the advantages of immobilised enzymes.

Importance of immobilised enzymes

Immobilised enzymes are enzymes that are fixed to an inert matrix. This can be achieved in TWO main ways:

1. Entrapment – held inside a gel, e.g. silica gel.

2. Micro-encapsulation – trapped inside a micro-capsule, e.g. alginate beads.

Beads containing the enzyme can be packed into a glass column, and substrate added at one end. The rate of flow of the substrate over the beads can be controlled: a slow flow rate will give more time for enzyme–substrate complexes to form, and therefore yield more product. Because the enzymes are contained within their own 'micro-environment', the enzymes are less susceptible to changes in pH, temperature and the action of chemicals such as organic solvents.

There are a number of advantages to immobilising enzymes in this way:

1. Enzyme can be easily recovered and reused.

2. Product is not contaminated by the enzyme.

3. More stable at higher temperature.

4. Catalyse reactions in a wider range of pH.

The result is that several enzymes with different temperature and pH optima can be used at the same time. Enzymes can also be easily added or removed giving greater control over the reaction.

Grade boost
When describing the advantages of immobilised enzymes in terms of activity over a range of pH it is important to say wider range, not just wide range.

quickfire

⑤ Describe three advantages of testing blood glucose using a biosensor over the Benedict's test.

≫ Pointer

Immobilised antibodies and enzymes are also commonly used to detect minute traces of the pregnancy hormone HCG in urine, and so confirm whether a patient is pregnant or not.

Biosensors

Biosensors contain immobilised enzymes that can be used to detect small concentrations of specific molecules in a mixture, e.g. glucose in a sample of blood. A biosensor consists of a specific immobilised enzyme, a selectively permeable membrane, and a transducer connected to a display. The selectively permeable membrane allows the metabolite to diffuse through to the immobilised enzyme, whilst preventing the passage of other molecules. The metabolite binds to the active site of the enzyme, and is converted into a product, which in turn combines with the transducer turning the chemical energy into an electrical signal. The higher the concentration of metabolite present, the greater the electrical signal. This technique is used to accurately measure the blood glucose of diabetic patients whose blood glucose should normally be kept between 3.89 and 5.83 mmol dm^{-3}.

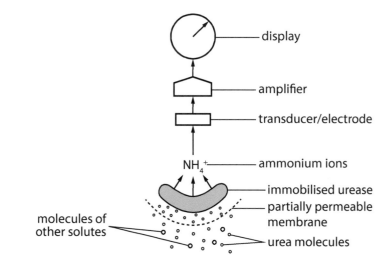

Biosensor

1.5 Nucleic acids and their functions

Nucleotides

Both DNA and RNA are made up of monomers called nucleotides: each nucleotide contains a phosphate group, a nitrogen-containing organic base, and a pentose (5-carbon) sugar: either ribose (RNA) or deoxyribose (DNA). There are two groups of organic bases: pyrimidines (single ring) and purines (double ring).

Four nitrogenous bases found in DNA:

- guanine (purine)
- cytosine (pyrimidine)
- adenine (purine)
- thymine (pyrimidine).

In RNA the pyrimidine uracil replaces thymine.

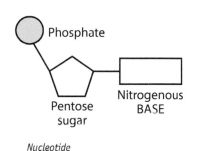

Nucleotide

Grade boost

Watch your spelling here. Candidates often write thiamine instead of thymine and cysteine instead of cytosine. Thiamine is a vitamin, and cysteine is an amino acid so you won't get the mark.

ATP

Adenosine **tri**phosphate is also a nucleotide: it has a ribose sugar joined to the adenine base, with three phosphate groups attached.

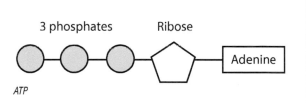

ATP

When the high-energy bond between the second and third phosphate group is broken via hydrolysis by the enzyme ATPase, 30.6kJ of energy is released for use in the cell, and adenosine diphosphate is formed. This reaction is reversible, requiring energy from respiration of glucose to reform the bond.

ATP → ADP + Pi + 30.6kJ energy
(Pi = inorganic phosphate)

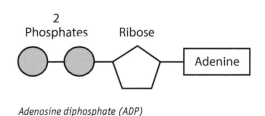

Adenosine diphosphate (ADP)

Grade boost

Watch your spelling – it is NOT <u>adenine</u> triphosphate.

>> Pointer

We have a total of 5g of ATP in our cells, but use up 50kg every day. This means that this reversible reaction must occur over 10,000 times a day because we cannot store it.

Grade boost

Be careful with roles and advantages of ATP. They are not the same thing!

Advantages of ATP:

- Energy is released quickly from a one-step reaction involving just one enzyme (hydrolysis of glucose takes many steps).
- Energy is released in small amounts, 30.6kJ where it is needed. By contrast just one molecule of glucose contains 1880kJ which couldn't safely be released all at once.
- It is the 'universal energy currency', i.e. it's a common source of energy for all reactions in all living things.

Roles of ATP in cells:

- used in many anabolic reactions, e.g. DNA and protein synthesis
- active transport
- muscle contraction
- nerve impulse transmission.

Structure of DNA

Two scientists, Watson and Crick, proposed the molecular structure of DNA in 1953, using information from other scientists including Franklin and Wilkins.

DNA consists of two polynucleotide strands that are arranged into a double helix. First a dinucleotide is formed when a condensation reaction occurs between two nucleotides: The 5th carbon atom of a deoxyribose sugar is joined to the 3rd carbon atom of the deoxyribose sugar of the nucleotide above it, via the phosphate molecule. This continues, building a single strand of DNA in the 5'-3' direction.

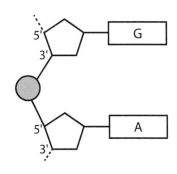

DNA dinucleotide

DNA then forms a double-stranded molecule from two strands: one strand runs in the opposite direction to the other (anti-parallel). Both strands are held together by hydrogen bonds that form between complementary nitrogenous bases. The double strand then twists to form a double helix.

Bases between both strands pair up in a certain way which is called the complementary base pairing rule: guanine forms hydrogen bonds with an adjacent cytosine molecule and adenine forms hydrogen bonds with an adjacent thymine molecule.

Hydrogen bonds are weak, but the sheer number of them present in a molecule of DNA over a million nucleotides long, means that collectively they are very strong. In fact you would need to heat DNA to over 95°C to break them all!

Pointer

Remember the rule G = C, A = T.

Pointer

Calculating proportion of bases present in DNA:

If you know that 22% of the DNA molecule is adenine, you can work out the proportions of the other bases. How? Following base pairing rule, A = T so 22% must also be thymine. The remainder 56% must be G + C. Therefore guanine = 28%, cytosine = 28%. NB this only works for double-stranded molecules.

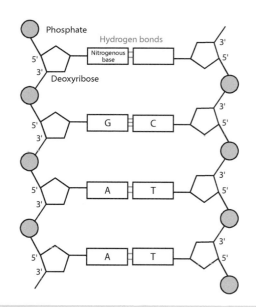

Phosphate

Hydrogen bonds

Nitrogenous base

Deoxyribose

Double stranded
DNA molecule

 Pointer

There are two hydrogen bonds between A and T (remember **A T**ea for two!), but three between G and C.

Pointer

The double stranded DNA molecule twists to form a double helix.

Extracting DNA

DNA can be easily extracted from cells by grinding up a sample in a solution of ice cold salt and washing up liquid. The detergent dissolves the lipids in the phospholipid membranes, allowing DNA to be released, and the cold temperature protects the DNA from cellular DNAases. Addition of protease will digest any remaining cellular enzymes and the histones that the DNA is wound around. Finally, adding ethanol to the salt already present, will cause the DNA to precipitate out from the solution.

Grade boost

When comparing DNA and mRNA remember that DNA is a double helix and mRNA is a single strand.

Structure of RNA

RNA is different from DNA, as it is usually a shorter, single-stranded molecule. Nucleotides also differ in that the sugar is ribose, and one base thymine is replaced with uracil. Three different types of RNA are involved in protein synthesis.

Molecule	Detail
mRNA	Messenger RNA is a single-stranded molecule typically 300–2000 nucleotides long. It is produced in the nucleus using one of the DNA strands as a template during transcription.
rRNA	Ribosomal RNA forms ribosomes with the addition of protein
tRNA	Transfer RNA is a small molecule that winds itself into a cloverleaf shape. It has an anticodon at one end, and an amino acid at the other. As the name suggests, it 'transfers' the correct amino acid to the growing polypeptide during translation.

Table showing the three different RNA molecules present

quickꟼire

① If a sample of nucleic acid contains 23% adenine, and 30% thymine, what can you deduce about the structure of the nucleic acid?

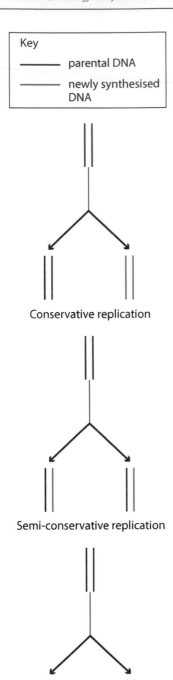

Key

—— parental DNA

—— newly synthesised DNA

Conservative replication

Semi-conservative replication

Dispersive replication

DNA replication theories

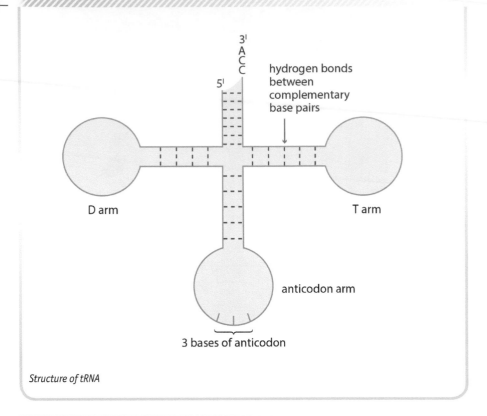

Structure of tRNA

Functions of DNA

DNA has two main functions in organisms:

1. Protein synthesis – the sequence of bases in one strand, called the template strand, determines the order of amino acids in the polypeptide (primary structure).

2. Replication – when cells divide, a complete copy of the DNA in the cell needs to be made. Both DNA strands separate, and each strand acts as a template to synthesise a complementary strand.

Three theories for how DNA replicates have been proposed:

1. Conservative replication: original parent double-stranded molecule is conserved, and a new double-stranded DNA molecule synthesised from it.

2. Semi-conservative replication: parental strands separate, and each strand acts as a template to synthesise a new strand. The new molecule consists of one original parent strand and one newly synthesised strand.

3. Dispersive: the newly synthesised molecules contain fragments from the original parent strand and newly synthesised DNA.

Process of semi-conservative DNA replication

The process requires ATP, free nucleotides, and enzymes.

- DNA helicase breaks the hydrogen bonds between the bases causing the double helix to unwind and separate into two strands.
- The exposed bases bind to free floating nucleotides in the nucleoplasm.
- DNA polymerase binds the complementary nucleotides (forming the phosphodiester bond).
- One strand acts as the template for the new molecule, so newly synthesised DNA contains one parent strand and a complementary newly synthesised strand.

Meselson–Stahl experiment

This was an important experiment carried out to determine the exact mechanism for DNA replication.

The experiment involved:

1. Growing bacteria on a ^{15}N medium. ^{15}N is a heavy isotope of nitrogen so all DNA produced would be of a heavier weight than normal. When DNA was extracted by centrifuging in caesium chloride, the DNA band appeared low down in the tube.

2. Bacteria were then grown on a ^{14}N medium (normal weight nitrogen), and after one generation the DNA extracted formed an intermediate band half way up the tube. This is because the DNA molecule contained one strand from the heavy parent DNA and one newly synthesised light DNA strand. (Because one band only was produced this rules out conservative replication.)

3. The bacteria were grown for a further generation using ^{14}N medium. The DNA extracted formed an intermediate band half way up the tube, and a lighter band towards the top of the tube. Because half of the DNA was intermediate weight and half light, this rules out dispersive replication.

4. DNA therefore replicates semi-conservatively.

5. If grown for further generations using ^{14}N medium, whilst intermediate weight DNA would remain, the proportion of light DNA produced would increase.

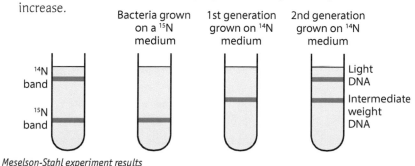

Meselson-Stahl experiment results

Grade boost

DNA polymerase does NOT form hydrogen bonds. It binds the complementary nucleotides (forming the phosphodiester bond).

 Pointer

DNA can be separated according to size by centrifuging samples at very high speeds (over 100,000 rpm) in a dense solution of caesium chloride. DNA will settle out where the density of the DNA equals that of the caesium chloride, and a visible band will form.

>> *Pointer*

Scientists first noticed that there were three times the number of bases in the DNA than amino acids in the polypeptide that it coded for.

Grade boost

You do not need to learn which codons code for each amino acid, just be able to use any codon table provided.

quickpire

② Convert the following DNA sequence into mRNA: GATTTCCGAATTGGCC.

Key Term

Codon: the triplet of bases in mRNA that codes for a particular amino acid, or a punctuation signal.

Grade boost

Make sure you are able to work out the sequence of amino acids for a DNA sequence using an mRNA codon.

The genetic code

The sequence of nucleotide bases forms a code. Each 'code word' has three letters called a triplet code or **codon**, which codes for a specific amino acid. The table shows a few examples:

DNA codon	mRNA codon	Amino acid that is coded for	Amino acid abbreviation
GGG	CCC	proline	Pro
CGG	GCC	glycine	Gly
ATG	UAC	tyrosine	tyr
TAC	AUG	methionine	met
ACT	UGA	stop (no amino acid)	

There are 20 amino acids that are coded by 4^3 bases, i.e. 64 different combinations of A, G, C, T(U). Therefore, there are 'spare' base codes. This is referred to as degeneracy or the 'degenerate code'.

This code is universal, i.e. it is the same in all living things. One codon acts as a START codon, marking the point on the DNA where transcription begins – this is AUG on the mRNA and codes for methionine. Each gene found on the DNA will code for a different polypeptide: this is called the one gene, one polypeptide hypothesis.

Protein synthesis

- Transcription occurs in the nucleus.
- Translation occurs at the ribosomes.
- Post-translational modification occurs in the Golgi apparatus prior to packaging of the protein into vesicles.

Transcription

- DNA acts as a template for the production of mRNA.
- DNA helicase acts on a specific region of the DNA molecule called the cistron, breaking the hydrogen bonds between both DNA strands, causing the strands to separate and unwind, exposing nucleotide bases.
- Free RNA nucleotides pair to exposed bases on the DNA template strand and RNA polymerase joins them by forming the phosphodiester bonds between the phosphate group on one nucleotide and the ribose sugar on the next.
- This continues until the RNA polymerase reaches a STOP codon, when the RNA polymerase detaches and production of mRNA is complete.
- The mRNA strand leaves the nucleus via the nuclear pores and moves to the ribosomes.

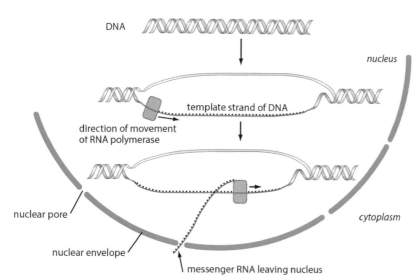

Transcription

In eukaryotes, **introns** are present within many genes so are also transcribed producing pre-mRNA. The coding regions are referred to as **exons**. The pre-mRNA is spliced to remove the non-coding regions *before* passing to the ribosomes. In prokaryotes, the DNA does not contain introns, and so the mRNA is produced directly from the DNA template.

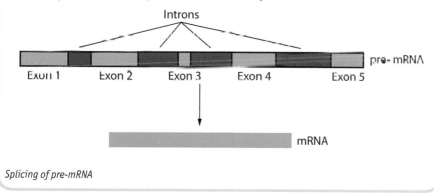

Splicing of pre-mRNA

Translation

Translation involves another specific RNA molecule called transfer RNA (tRNA). At one end of the tRNA molecule there are three exposed bases called the anticodon, these are complementary to the mRNA codon. At the opposite end of the tRNA molecule is an amino acid attachment site where the relevant amino acid is found. The attachment of the relevant amino acid to the attachment site is called amino acid activation and requires ATP.

Translation involves converting the codons on the mRNA into a sequence of amino acids known as a polypeptide. Each ribosome (found free in the cytoplasm, or attached to the rough endoplasmic reticulum – see page 24) is made up of two subunits made from ribosomal RNA and protein. The mRNA binds to the smaller subunit, whilst tRNA binds to one of two attachment sites on the larger subunit.

Grade boost

Don't get DNA polymerase and RNA polymerase mixed up!

Key Terms

Introns: non-coding nucleotide sequence in DNA and pre-mRNA, that is removed from pre-mRNA, to produce mature mRNA.

Exons: nucleotide sequence on one strand of the DNA molecule and the corresponding mRNA that codes for the production of a specific polypeptide.

quickfire

③ Name the enzyme involved in the production of mRNA.

quickfire

④ Explain why pre-mRNA is spliced in eukaryotes.

The process involves a number of steps:

- Initiation: ribosome attaches to the START codon.
- tRNA molecule with a complementary anticodon to the first codon, binds to the first attachment site on the ribosome.
- A second tRNA molecule joins to the second attachment site, and a ribosomal enzyme catalyses the formation of a peptide bond between the two amino acids. This is known as elongation.
- The first tRNA molecule is released and the ribosome now moves one codon along the mRNA, which exposes a free attachment site and another tRNA molecule joins and the process is repeated.
- This repeats until a STOP codon is reached, when the polypeptide is released. This is called termination.
- Usually several ribosomes bind to a single mRNA strand at the same time. This is called a polysome.

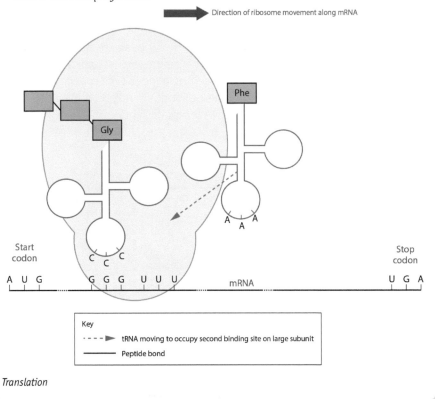

Translation

Post-translational modification

Translation produces a polypeptide, but further modification is needed in order to produce a protein with a secondary, tertiary or quaternary structure. This modification occurs within the Golgi body. Modification also occurs to produce molecules such as glycoproteins, lipoproteins, and complex quaternary structures such as haemoglobin. To form haemoglobin, two alpha chains and two beta chains (coded by two different genes) need to be assembled together with iron as a prosthetic group.

1.6 The cell cycle and cell division

Chromosomes

Chromosomes consist of DNA and a protein called histone, and are only visible after they condense at the onset of cell division. Following DNA replication, a chromosome exists as two identical 'sister' chromatids joined by the centromere. Sister chromatids are genetically *identical*. Each chromosome contains **genes** that code for specific polypeptides.

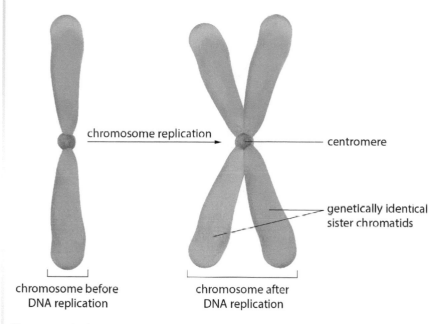

chromosome replication

centromere

genetically identical sister chromatids

chromosome before DNA replication

chromosome after DNA replication

Chromosome structure

Different species have different numbers of chromosomes e.g. humans have 46, whilst a potato has 48. In humans, chromosomes come in 23 pairs: one from each parent. It is these pairs that are said to be **homologous**, i.e. they contain the same genes but they may be different versions or **alleles**. Where an organism is said to have two complete sets of chromosomes this is called diploid, so in humans the diploid number (2n) is 46. The potato has four complete sets of chromosomes – this type of **polyploidy** is referred to as tetraploid (4n). Haploid (n) numbers are found in human gametes and also in some organisms, e.g. mosses, male worker bees.

Key Terms

Gene: a base sequence of DNA that codes for the amino acid sequence of a polypeptide. Each gene occupies a fixed position on the chromosome called the locus.

Homologous chromosomes: homologous chromosomes are the same size and shape and carry the same genes but these may be different versions called alleles. One chromosome of each pair comes from each parent.

Allele: a different form of the same gene.

Polyploidy: where an organism has more than two complete sets of chromosomes.

Cell cycle

The majority of the cell cycle in eukaryotes involves **interphase**, where DNA, protein and organelles are synthesised. This is followed by four stages of **mitosis** where chromosomes are arranged and separated, prior to the formation of two genetically identical cells via **cytokinesis**.

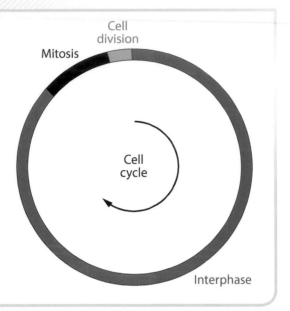

Cell cycle

Interphase

This is the longest phase in the cell cycle, and is very metabolically active. The quantity of DNA doubles (though the chromosome number actually remains the same, as chromosomes exist as two sister chromatids joined together at the centromere – see diagram above), and protein synthesis and organelle replication occur, requiring much ATP. The cell is metabolically very active.

Mitosis

Mitosis results in the production of two genetically identical cells. It is important in growth and repair when differentiated cells replicate. It consists of four stages:

Stage	What happens	Diagram
Prophase	• Chromosomes condense to become shorter and thicker • Chromosomes become visible as two sister chromatids attached by a centromere • Centrioles move to opposite poles (not higher plants) • Nuclear envelope disintegrates • Nucleolus disappears	**Early prophase** nuclear envelope centrioles cytoplasm nucleolus cell membrane chromosomes condensing **Late prophase** nuclear envelope disintegrates nucleolus disappears centromere centrioles moving to opposite ends (poles) of cell pair of chromatids
Metaphase	• Spindle forms • Chromosomes align at the equator of the cell attached to spindle microtubules via centromere	**Metaphase** each centriole reaches a pole; they organise production of the spindle microtubules spindle chromosomes line up across the equator of the spindle attached by their centromeres to the spindle
Anaphase	• Spindle fibres shorten • Centromeres divide, and chromatids are pulled towards opposite poles	**Anaphase** centrioles chromatids move to opposite poles; centromeres first, pulled by the microtubules
Telophase	• Chromatids reach poles and become indistinct by uncoiling • Nuclear envelope reforms • Nucleolus reforms • Spindle disintegrates	**Telophase** cleavage furrow nuclear envelope re-forming cell membrane centrioles nucleolus re-forming remains of spindle which is breaking down chromatids have reached the poles of the spindle

Grade boost

Try to remember PMAT = prophase (prominent DNA), metaphase (middle), anaphase (apart) and telophase (two).

Grade boost

Be prepared to draw or identify chromosomes in any of the four stages of mitosis.

quickfire

① Match the following events to the correct stage in mitosis. You may use a letter once, more than once, or not at all.

1. Chromosomes align at the equator
2. Nuclear envelope disappears
3. Cleavage furrow seen
4. Chromatids move to opposite poles
5. Spindle fibres shorten

A Prophase
B Metaphase
C Anaphase
D Telophase

Cytokinesis

The division of the cytoplasm into two distinct cells is different in plants and animal cells. In animal cells, the membrane infolds via a cleavage furrow, until the two cells become separated. In plant cells, the presence of the cellulose cell wall prevents this from happening, so instead a cell plate develops from the centre outwards, until the cell is divided into two.

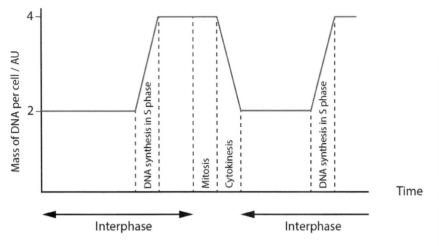

Graph showing changes in the mass of DNA in a cell during the cell cycle

>> **Pointer**

Calculating the percentage of cells undergoing mitosis = (number of cells in prophase + metaphase + anaphase + telophase / total number of cells) × 100.

quickfire

② Put the following cell drawings of the stages in mitosis into the correct order.

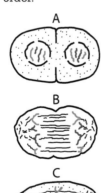

Significance of mitosis

By producing new cells, an organism can grow, repair tissues and replace dead cells. In animals, skin and blood cells are constantly being replaced as they wear out. In plants, cells at the root and shoot tip meristems are constantly undergoing mitosis.

The mitotic index is defined as the ratio of the number of cells in a population undergoing mitosis to the number of cells not undergoing mitosis, and is a measure of growth.

Mitosis is important in asexual reproduction where genetically identical offspring can be produced resulting in a rapid increase in numbers during favourable conditions, e.g. yeast cells and bacteria. It is also used by some plants, e.g. strawberries, when runners are produced.

Mitosis is controlled by a number of genes including proto-oncogenes. A mutation in one of these genes from chemicals, e.g. benzene, or radiation, e.g. UV light, will cause them to turn them into oncogenes: this results in uncontrollable cell division, which leads to the formation of tumours and cancers. Vincristine has been a successful drug in the treatment of cancers as it prevents the formation of the spindle so arresting mitosis at metaphase, slowing the rate of cell division.

Calculating the length of a stage in the cell cycle

1. First calculate the proportion of cells in that stage by counting using a microscope, e.g. 20 cells seen, with 16 in interphase (16/20 = 80%).
2. Apply the proportion to the length of the cell cycle, e.g. 24-hour cycle × 80% = 19.2 hours (19 hours and 12 mins) is the length of interphase.

Meiosis

Meiosis involves two consecutive cell divisions, and unlike mitosis, produces four genetically different, haploid cells. It occurs in the reproductive organs of animals, plants and some protoctistans prior to sexual reproduction. Interphase only occurs before prophase 1, and is responsible for DNA replication and protein synthesis.

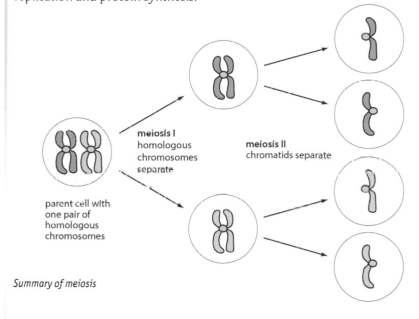

Summary of meiosis

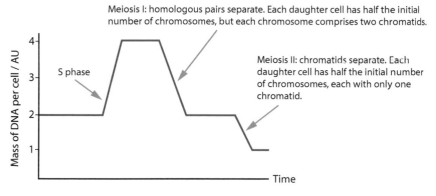

Graph showing changes in the mass of DNA in a cell during meiosis

>> *Pointer*
Mitosis and meiosis are both continuous processes. Under the microscope you only see a snap shot of what is happening.

quickfire

③ Interphase occurs in both mitosis and meiosis and is a period of intense activity. State three events that occur.

Grade boost
Watch your spelling here, as mitosis and meiosis can be easily confused. For example, meitosis = 0 marks!

Meiosis 1

Following interphase, meiosis 1 follows a series of steps similar to mitosis, except there is a modified prophase 1 when homologous pairs come together to form bivalents and crossing over may occur. This increases genetic variation. The other major difference occurs during metaphase 1, when bivalents align randomly at the equator. This is known as independent assortment and gives a further 8388608 variants from 23 bivalents i.e. 2^{23}.

Stage	What happens	DNA content (a.u.)	Chromosome number per cell
Interphase	Occurs before meiosis DNA replicated	2	2n
Prophase 1 *Key difference = crossing over can occur*	Chromosomes condense to become shorter and thicker Centrioles move to opposite poles (not higher plants) Chromosomes come together in homologous pairs (bivalent) Crossing over occurs – part of one chromatid is exchanged with another Nucleolus and nuclear membrane disappear	4	2n
Metaphase 1 *Key difference = bivalents align*	Spindle forms Homologous chromosome pairs (bivalents) align at the equator of the cell attached to spindle microtubules via centromere. This alignment is random and called independent assortment	4	2n
Anaphase 1 *Key difference = chromosomes are pulled to opposite poles*	Spindle fibres shorten Bivalents separate and chromosomes are pulled towards opposite poles	4	2n
Telophase 1	The chromosomes reach poles In some cases: Nuclear envelope reforms Nucleolus reforms Spindle disintegrates	4	2n
Cytokinesis	Division of cytoplasm occurs, creating two haploid cells	2	n

Meiosis 2

Similar to mitosis as there is no pairing of homologous chromosomes.

Prophase 2	Centrioles separate, and arrange themselves at 90° to the previous spindle	2	n
Metaphase 2	Chromosomes align at the equator of the cell attached to spindle microtubules via centromere	2	n
Anaphase 2	Spindle fibres shorten Centromeres divide, and chromatids are pulled towards opposite poles	2	n
Telophase 2	Chromatids reach poles and become indistinct Nuclear envelope reforms Nucleolus reforms Spindle disintegrates	2	n
Cytokinesis	Four haploid daughter cells produced	1	n

Significance of meiosis

- Generates genetic variation through crossing over (prophase 1) and independent assortment (metaphase 1 and 2).
- Keeps the chromosome number constant: by producing haploid gametes that recombine during fertilisation, restoring the diploid number in the zygote.

Comparison of mitosis and meiosis

Mitosis	Meiosis
One cell division	Two cell divisions
Produces genetically identical cells	Produces genetically different cells
Cells are diploid	Cells are haploid
No crossing over	Crossing over occurs in prophase 1
No independent assortment	Independent assortment occurs in metaphase 1 and 2

Grade boost

When trying to identify cells in meiosis 1 or 2, a good guide is to count the number of cells – two cells usually indicates meiosis 2.

quickfire

④ Match the following events to the correct stage in meiosis. You may use a letter once, more than once, or not at all.

1 Bivalents form.
2 Nuclear envelope disappears.
3 Independent assortment occurs.
4 Chromatids move to opposite poles.
5 Spindle fibres shorten.

A Prophase 1
B Metaphase 1
C Anaphase 2
D Telophase
E Anaphase 1

quickfire

⑤ Identify the following stages in meiosis.

Component 1 Summary

Chemical elements and biological compounds

Properties linked to the chemical structure of:

- Inorganic ions – magnesium (chlorophyll), iron (haemoglobin), phosphate ions (ATP, DNA, RNA)
- Water – important solvent and involved in many biochemical reactions
- Building polymers via condensation reactions and breaking down polymers via hydrolysis
- Carbohydrates – source of energy, some polymers act as energy storage molecules, e.g. starch and glycogen, whilst other polymers add strength and support, e.g. chitin and cellulose
- Lipids – twice energy than carbohydrates, provide both thermal and electrical insulation, form part of cell membrane and provides protection against physical damage around organs, e.g. kidney
- Proteins – function as enzymes, hormones, antibodies, transport and structural component of cell membranes

Cell structure and organisation

- Cell structure – eukaryotes: structure and function of organelles. Be able to identify different organelles from electron micrographs. Prokaryotes are simple bacteria that lack organelles. Viruses cause a variety of infectious diseases in plants and animals
- Main organelles are 1) nucleus the site of transcription, 2) mitochondria the site of aerobic respiration, 3) chloroplasts the site of photosynthesis, 4) ribosomes the site of translation and 5) Golgi body where proteins are modified and packaged for export from the cell
- Levels of organisation – tissues and organs
- Tissue types including connective, muscular and epithelial

Cell membranes and transport

- Cell membrane – phospholipid bilayer and fluid mosaic model
- Transport across membrane via: diffusion, facilitated diffusion, active transport, co-transport, osmosis and via bulk transport.
- Effect of temperature and organic solvents on membrane permeability

Enzymes and biological reactions

- Enzyme structure – primary (order of amino acids), secondary (alpha helix and beta pleated sheet), tertiary (folding of secondary structure to give 3D shape held together with hydrogen, ionic and disulphide bonds), and quaternary structure (two or more polypeptides joined together)
- How enzymes work via lock and key and induced fit hypotheses
- Factors affecting enzyme action – temperature, pH, concentration of substrate and enzyme
- Enzyme inhibition – competitive and non-competitive and end product inhibition
- Immobilised enzymes – benefits and uses in industry

The cell cycle and cell division

- Mitosis and cell cycle – interphase, prophase, metaphase, anaphase and telophase, producing genetically identical daughter cells during growth and repair
- Meiosis producing haploid, genetically different gametes during sexual reproduction

Nucleic acids and their functions

- The differences between the structures of ATP, DNA and RNA
- DNA replication – role of DNA polymerase during cell division and how Meselson-Stahl experiment provides evidence for semi-conservative DNA replication
- Protein synthesis – role of RNA polymerase in transcription and ribosomes in translation, and Golgi body in modification and packaging of proteins.

Knowledge and Understanding

Component 2

Classification and biodiversity
p70–77

Adaptations for gas exchange
p78–85

Adaptations for transport in animals
p86–94

Biodiversity and physiology of body systems

Adaptations for transport in plants
p95–103

Adaptations for nutrition
p104–115

Basic notes   Good grasp   Fully revised

Classification and biodiversity

Living organisms are classified into a taxonomic hierarchy based upon their physical features. This topic looks at the different methods that are used to assess the relatedness of organisms. Biodiversity has been generated through the adaptation of anatomical, physiological and behavioural traits of organisms through natural selection.

→ **p70–77** → ☐ ☐ ☐

Aptations for gas exchange

Different organisms use a variety of methods to exchange gases with their environments. This topic looks at the structure and functioning of gas exchange surfaces in unicellular and multicellular plants and animals, and how they are adapted to their specific environments.

→ **p78–85** → ☐ ☐ ☐

Adaptations for transport in animals

Multicellular organisms need a transport system to carry materials from where they are taken in from the environment, to all cells in the body and to remove waste products. This topic looks at the different transport systems in organisms, and the role of blood and the cardiovascular system in mammals.

→ **p86–94** → ☐ ☐ ☐

Adaptations for transport in plants

Plants must absorb water and minerals from the soil and transport it to the leaves where it is used in photosynthesis, and then transport the products of photosynthesis to the growth and storage regions of the plant. This topic covers the vascular system in plants and the mechanisms by which water and solutes are transported.

→ **p95–103** → ☐ ☐ ☐

Adaptations for nutrition

Living organisms need nutrients to provide energy for growth and internal processes. This topic covers the different forms of nutrition, how humans digest food and how other organisms are adapted to their different diets.

→ **p104–115** → ☐ ☐ ☐

2.1 Classification and biodiversity

Classification

Biological classification is said to be phylogenetic: it reflects the evolution of an organism, by placing organisms into groups according to visible external features. A phylogenetic tree shows how organisms are related by showing their common ancestors.

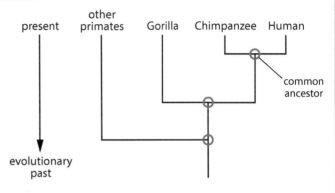

Phylogenetic tree

Classification uses a hierarchy whereby smaller groups are placed into larger ones, with no overlaps between the groups, called taxa. One example is: domain, kingdom, phylum, class, order, family, genus, species.

1. There are three **domains**: Archaea (bacteria living in hostile environments, e.g. extreme temperature, pH, salinity or pressure, with an unusual metabolism, e.g. producing methane), Eubacteria (common bacteria) and Eukarya (the eukaryotes – includes plants, animals, fungi and protoctists). This is the highest category into which organisms are classified.

2. Organisms are then classified into five kingdoms based upon their physical characteristics: Plantae (plants), Animalia (animals), Fungi, Prokaryotes (bacteria), and Protoctista. Protoctists are mainly unicellular eukaryotic microorganisms that do not form tissues. Many are photosynthetic, e.g. algae.

3. The kingdoms are then sorted into a large number of smaller groups called phyla. All members of a given phylum or division have certain things in common, e.g. the chordates all have a spinal cord.

4. Each class that is a sub-group of a phylum, e.g. mammals are a class within chordates, is subdivided into orders, e.g. humans belongs to the order of primates.

5. The subdivision of an order contains different families.

6. Families are subdivided into genera (singular genus). A genus is a group of organisms with a large number of similarities but members of different species within a genus are usually unable to interbreed successfully, e.g. the horse and the donkey can produce offspring in the form of the mule, but the mule itself is sterile.

>> *Pointer*

A good way to remember the order of these is using the rhyme: Did King Phillip Come Over From Germany Swimming.

Key Term

Domains: contain organisms that share a distinctive, unique pattern of ribosomal RNA, which establishes their close evolutionary relationship.

7. Finally genera are subdivided into **species** – a category that consists of a group of similar individuals that can interbreed and produce fertile offspring. They have a very large number of anatomical and physiological similarities.

Binomial system

Every organism has two names. This naming system was first introduced by a Swedish scientist, Carl Linnaeus, in 1735. The first name used is the name of the genus to which the organism belongs. The genus name has an initial capital letter, e.g. *Homo* (man). The second name used is the name of the species to which the organism belongs, and this name is possessed by only one kind of organism, and has a small initial letter, e.g. *sapiens* (modern). The use of the binomial system is universal and helps avoid confusion between different languages.

By classifying organisms we can infer evolutionary relationships, and it makes it easier to manage the large number of organisms. It is, however, tentative and may be subject to change as new species are discovered which do not fit neatly into the groups currently available.

Five kingdom system

Kingdom	Key features
Plantae	Are multicellular eukaryotic organisms that photosynthesise (**autotrophic**). Reproduce using spores (e.g. mosses and ferns) or seeds (e.g. flowering plants and conifers). Possess cellulose cell walls.
Animalia	Are multicellular, **heterotrophic**, eukaryotic organisms. Lack cell walls. Show nervous coordination.
Fungi	Are multicellular (e.g. moulds) or single celled (e.g. yeast), eukaryotic organisms. In moulds, the body consists of a network of threads called hyphae. Cell wall is made of chitin. They are heterotrophic being either **saprophytic** or **parasitic**. Reproduce by production of spores (moulds) or by budding (e.g. yeast).
Prokaryotes	Microscopic, unicellular organisms including bacteria and cyanobacteria (blue-green algae). Cell wall is made of peptidoglycan (murein). Lacks membrane-bound organelles, and a true nucleus. Ribosomes are smaller than eukaryotes (70S).
Protoctista	Includes algae and slime moulds. Some are unicellular and resemble animal cells (e.g. *Amoeba*) whilst others are colonial and have plant like-cells (e.g. *Spirogyra*). Contain membrane-bound organelles and a nucleus.

Key Terms

Species: consists of a group of individuals with similar characteristics that can *interbreed* **and** *produce fertile offspring.*

Autotrophic nutrition: is making complex organic molecules from simple inorganic ones using either light or chemical energy.

Heterotrophic nutrition: involves consumption of already made complex organic molecules.

Saprophytic nutrition: involves feeding on dead or decaying matter by the production of enzymes extracellularly and the subsequent absorption of the products.

Parasitic nutrition: involves obtaining nutrients from a host organism over a long period of time, causing it harm in the process.

 Grade boost

You should be able to list key features of each kingdom.

Assessing relatedness

Determining the relatedness of species was initially performed by looking at physical characteristics from living organisms and fossil evidence. The use of immunology and more recently DNA profiling has led to a greater understanding of how closely related organisms are.

Comparing physical features

When comparing features, taxonomists look for similar or **homologous structures**. An example in vertebrates is the pentadactyl limb (which means having five digits). Its basic structure is similar in amphibians, reptiles, birds and mammals, though they have very different functions in each. This shows evidence of divergent evolution where a structure has evolved from a common ancestor to perform a different function.

Vertebrate fore limb	Diagram	Function
Human		Grasping
Bird		Flight
Whale		Swimming

Key Terms

Homologous: structures in different species with a similar anatomical position and developmental origin, derived from a common ancestor.

Analogous structures: have a corresponding function and similar shape, but have a different developmental origin.

quickfire

① Using examples, describe the difference between analogous and homologous structures.

Where **structures** are **analogous**, i.e. they have the same function but a very different shape/structure; this provides evidence for convergent evolution whereby ancestors have adapted themselves to the same environmental pressure but from different developmental origins. An example is found with the wings of a bird and those of a butterfly: both provide flight, but due to their very different structures, there is no evidence that they shared a common ancestor and so wings should not be used to classify them.

Immunology

The main immunological technique used to show relatedness of species is the immunological comparison of proteins, and involves creating antibodies to one species' protein in a rabbit, which can then be presented to other species' proteins. For example, if you wanted to compare living primates with a human to find its closest ancestor, a protein that is present in all species would first need to be identified. The human protein is then injected into a rabbit to produce antibodies to it. This antibody serum is then added to protein from the other primates, e.g. chimpanzee, gibbon, and gorilla, and the degree of precipitation measured. If human antibody is added to the human protein, an exact antibody-antigen match occurs and there will be 100% precipitation. As the similarity between the proteins diminishes, a lower degree of precipitation will be observed. This has been done with haemoglobin, and the closest relatives were found to be the chimpanzee and gorilla both with 95% precipitation. When comparing the actual amino acid sequence of haemoglobin from gorillas and chimpanzees with humans, the closest match was with the chimpanzee.

DNA fingerprinting and sequencing

As species evolve, changes occur in their DNA base sequences, so more closely related organisms show fewer differences in their base sequences. Comparing genetic (DNA) fingerprints and DNA sequences of genes has confirmed evolutionary relationships, e.g. DNA sequencing has confirmed that Darwin's finches did indeed evolve from the mainland ancestral finch, as he had thought.

Based upon DNA sequence similarity, our closest ancestor is the chimpanzee, followed by gorilla then orang-utan.

In eukaryotes, the majority of the DNA does not code for polypeptides: these non-coding regions between genes contain short DNA sequences that repeat, e.g. TATATATATATATATA, and are called satellites or STRs (short tandem repeats). The number of times that these repeat is unique and so forms the basis of genetic fingerprinting.

 Pointer

The greater the variation in the base sequence, the greater the genetic diversity of the species.

Key Term

Biodiversity: the number of species and the number of individuals of each species in a specified geographic region.

Grade boost

As the biodiversity of plant species increases, it creates more potential habitats for other organisms, e.g. insects, so their biodiversity is likely to increase as well.

Biodiversity

Species richness is a measure of the number of different species in a community. This together with the number of organisms within each species represents the **biodiversity** in that geographical region. Therefore, a field that has an abundance of species all with healthy numbers has a greater diversity than a field with the same number of species but with very small numbers.

As you move from the poles towards the equator, the biodiversity increases. This is in part due to increasing light intensity, but water availability is also important: a hot desert has a lower biodiversity than a temperate forest. The higher light intensity ensures that the solar energy entering these ecosystems is higher, allowing greater photosynthesis, but in the desert, the lack of water will limit plant growth and hence biodiversity.

Factors affecting biodiversity

1. Succession: the composition of a community changes over time as different species colonise.
2. Natural selection: see page 77.
3. Human activity: pollution, overfishing, and deforestation have all taken their toll on biodiversity by the physical removal of species or destruction of their habitats. Farming has also played a part, as monoculture involves the growing of a single species of crop, e.g. maize, so other species are removed to maximise yields.

Reductions in biodiversity are a concern because many plant species provide staple foods, e.g. rice and wheat, and provide raw materials, e.g. cotton. Many drugs are derived from plants, e.g. vincristine, which is used to treat cancer, and many have yet to be discovered.

Assessing biodiversity

Pointer

The value of Simpson's diversity index is between 0 and 1. The closer the value is to 1, the greater the diversity. The index can be used to compare different ecosystems.

An index of diversity measures the number of individuals of each species and the number of species. The species diversity can be calculated using an index of diversity formula, such as the Simpson's diversity index:

$$S = 1 - \frac{\Sigma n(n-1)}{N(N-1)}$$

where : N is the total number of organisms of all species
n is the total number of organisms of each species
Σ means the sum of

Worked example:

Species	Number of plants per m² in field A (n)	n(n–1)
Buttercup	12	12 × 11 = 132
Daisy	8	8 × 7 = 56
Plantain	9	9 × 8 = 72
Clover	13	13 × 12 = 156
Thistle	12	12 × 11 = 132
Dandelion	11	11 × 10 = 110
Bracken	0	0
Nettle	0	0
	N = 65 (N–1) = 64	Σ n (n–1) = 658

Field A

$$S = 1 - \frac{658}{65 \times 64}$$

$$= 1 - 0.16 \text{ (2 d.p.)}$$

$$= 0.84$$

Polymorphic loci and biodiversity

A **gene**'s position on the chromosome is referred to as its locus (loci). A locus shows polymorphism if it has two or more **alleles** that cannot be accounted for by mutation alone, resulting in two or more different **phenotypes**. In ABO blood grouping, the gene responsible for producing antigens on the surface of red blood cells has three different alleles: A, B and O. In some parts of the world, the frequency of the O allele is very high, accounting for over 99% of the **gene pool**. In other countries, the proportion of A and B alleles is much higher: this represents a higher biodiversity.

>> *Pointer*

Think of a gene like a car, and the allele is the make, e.g. BMW, Ford, etc.

>> *Pointer*

Genetic biodiversity can be assessed by determining the number of alleles at a locus (e.g. three in the case of ABO blood group) and the proportion of the population that have a particular allele.

Different sampling techniques

To estimate the number of individuals of each species in a given area, a number of practical techniques can be used. Sampling should be at random to eliminate sampling bias.

quickfire

② If field A has a diversity of 0.84, and field B is 0.79, which field has the greatest biodiversity?

Grade boost

You could be asked to identify improvements in a method used to generate results, or to describe how you would generate valid data.

Population	Technique	Method
Terrestrial animals	Mark-release-recapture (Lincoln Index)	Animals are captured and marked (it is important that they are not harmed or made more visible to predators) and then released. Once animals have had a chance to reintegrate with the population, e.g. 24 hours, the traps are reset. The total population size can be estimated using the number of individuals captured in sample 2, and the number in that sample that are marked (i.e. caught before). $$\text{Pop size} = \frac{\text{no. in sample 1} \times \text{no. in sample 2}}{\text{no. marked in sample}}$$ Have to assume that no births/deaths/immigration/emigration have occurred during the time between collecting both samples.
Freshwater invertebrates	Use kick-sampling and use Simpson's Index	Collect and identify invertebrates from a given area using a quadrat and a net. Kick or rake the area, e.g. $0.5m^2$ for a set period, e.g. 30 seconds, and collect invertebrates in a net downstream. Release invertebrates carefully. Use Simpsons Index to calculate diversity.
Plants	Quadrats and transects	Estimate percentage area cover of different plants using a quadrat divided into 100 sections. Measure plant diversity by counting number of plants in a quadrat, e.g. $1m^2$. A transect is a length of rope that can be used to measure intervals along an environmental gradient, e.g. distance from a woodland, along which quadrats can be placed.

Natural selection

Evolution is the process by which new species are formed from pre-existing ones over a period of time. Darwin's observations of variation within a population led to the development of the idea of natural selection. Darwin recognised that species changed. He proposed the theory of natural selection to explain why it happened. Natural selection results in species that are better adapted to their environment.

Organisms overproduce offspring so that there is a large variation of genotypes within the population. Changes to environmental conditions bring new selection pressures through competition/predation/disease. Only those individuals with beneficial alleles have a selective advantage, e.g. white fur in arctic, and are therefore more likely to survive. These individuals then reproduce, so offspring are likely to inherit the beneficial alleles, therefore the beneficial allele frequency increases within the gene pool.

These adaptations may be:

1. Anatomical, e.g. beak shape in finches.
2. Physiological, e.g. haemoglobin with a higher affinity for oxygen as found in llamas that live at high altitudes.
3. Behavioural, e.g. nocturnal animals.

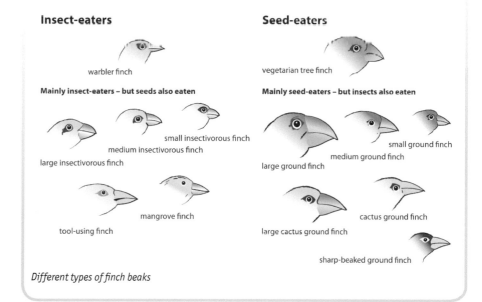

Insect-eaters

warbler finch

Mainly insect-eaters – but seeds also eaten

small insectivorous finch
medium insectivorous finch
large insectivorous finch

tool-using finch
mangrove finch

Seed-eaters

vegetarian tree finch

Mainly seed-eaters – but insects also eaten

small ground finch
medium ground finch
large ground finch

large cactus ground finch
cactus ground finch

sharp-beaked ground finch

Different types of finch beaks

Grade boost

It is important to use the term allele NOT gene when describing natural selection.

Pointer

You need to appreciate the role of selective predation in natural selection, but the detailed mechanism of evolution is not required at AS/Year1.

quickfire

③ True or false?

A An allele is the same form of a gene.

B Genetic diversity represents the number of different alleles of genes in a population.

C Random mutation can result in new alleles of a gene.

D Species richness is a measure of the number of different species in a community.

2.2 Adaptations for gas exchange

Surface area to volume ratio

Pointer

Larger organisms have a greater number of cells and therefore have higher oxygen requirements.

Organisms exchange gases such as oxygen and carbon dioxide with the atmosphere via a gas exchange surface. The surface area of this surface determines how much can be exchanged. When an organism doubles in size, its volume (and therefore also its oxygen requirements) is cubed, but the surface area is only squared. Therefore, as organisms increase in size, a specialised gas exchange surface is required to increase the area available. Since this also increases the area available for water loss, there is always a balance to be struck between exchanging gases and water loss in terrestrial organisms.

Surface area	$1 \times 1 \times 6$ sides $= 6mm^2$	$2 \times 2 \times 6$ sides $= 24mm^2$	$4 \times 4 \times 6$ sides $= 96mm^2$
Volume	$1 \times 1 \times 1 = 1mm^3$	$2 \times 2 \times 2 = 8mm^3$	$4 \times 4 \times 4 = 64mm^3$
Surface area: Volume ratio	$6 : 1$	$3 : 1$	$1.5 : 1$

Surface area to volume ratio

Grade boost

There are some additional features, but these are NOT present in all organisms, e.g.

- Good blood supply to maintain the concentration gradient (*not* single-celled organisms, insects or plants).
- Ventilation mechanism to maintain concentration gradient (*not* single-celled organisms, worms or plants).

General characteristics of a gas exchange surface

- Large surface area to volume ratio
- Moist to allow gases to dissolve
- Thin to provide a short diffusion distance
- Permeable to gases.

Unicellular organisms

In single-celled organisms, e.g. *Amoeba*, the surface area is large enough to meet the needs of the organism and therefore materials can be exchanged directly across its thin and permeable cell surface membrane. As the cytoplasm is constantly moving, the concentration gradient is always maintained.

Multicellular animals

In larger organisms the surface area to volume ratio decreases, so diffusion across the body surface is insufficient to meet the needs of the organism. A number of adaptations have evolved to solve these problems, which become more specialised the larger the organism. Where animals are very active and therefore have a higher metabolic rate, their oxygen requirements cannot be supplied by the body surface alone. This is often solved by the presence of a specialised gas exchange surface with a ventilation mechanism that ensures that the concentration gradient is maintained across the respiratory surface.

One consequence of maintaining a moist respiratory surface in terrestrial animals is water loss: this is minimised by having internal gas exchange surfaces, called lungs.

Organism	Adaptations
Flatworm	Has a flattened body to reduce the diffusion distance between the surface and the cells inside and to increase the overall surface area (we saw this as an adaptation to cylindrical shaped mitochondria see page 23).
Earthworm	Secretes mucus to maintain a moist surface and has a well developed capillary network under the skin. Has a low metabolic rate to reduce oxygen requirements. Has a network of blood vessels and blood containing haemoglobin for the transport of oxygen. Carbon dioxide is transported largely in the blood plasma.
Amphibians, e.g frogs and newts	Moist and permeable skin with a well developed capillary network beneath the surface Have lungs that are used when more active.
Reptiles, e.g. snakes and crocodiles	Have internal lungs like amphibians, but these are more complex and have a larger surface area.
Birds	Flight generates a very high metabolic rate and hence oxygen requirement. To meet this, birds have an efficient ventilation mechanism to increase concentration gradient across the lung surface.

Gas exchange in fish

Fish have developed a specialised internal gas exchange surface called gills that are made up of numerous gill filaments containing gill lamellae at right angles to the filaments. These greatly increase the surface area for the exchange of oxygen and carbon dioxide.

Fish ventilate their gills in two different ways:

1. Cartilaginous fish, e.g. the shark: blood and water flow in the same direction over the gill (**parallel flow**). Gas exchange is only possible over part of the gill filament surface as an equilibrium is reached which prevents further diffusion and reduces the oxygen that can be absorbed into the blood. The ventilation mechanism in cartilaginous fish is basic: as they swim, they open their mouth allowing water to pass over the gills.

Pointer

If you find the concept of surface area : volume ratio difficult, just remember that as an organism gets bigger whilst keeping the same shape, the distance to the centre of the organism increases.

quickfire

① Why do multicellular animals require specialised exchange surfaces?

quickfire

② Name three different gas exchange surfaces.

Key Term

Parallel flow: blood and water flow in the same direction at the gill lamellae, maintaining the concentration gradient for oxygen to diffuse into the blood only up to the point where its concentration in the blood and water is equal.

Water flow	100%	80%	60%	50%	50%	50%
diffusion ↓ O₂						
Blood flow	0%	20%	40%	50%	50%	50%

numbers are % saturation with oxygen

Parallel flow

2. **Counter-current flow** is seen in bony fish, e.g. salmon, where blood and water flow in opposite directions. This is a much more efficient system because diffusion is maintained along the entire length of the gill filament because there is always a higher concentration of oxygen in the water than in the blood it meets, which results in higher oxygen absorption as an equilibrium is not reached. Bony fish have a more advanced ventilation mechanism than cartilaginous fish.

water flow running in opposite direction to blood flow in capillaries of gill lamella

gill filament

gill arch / bar

Counter-current flow across gill lamellae

gill plate / lamella

% saturation with oxygen

water

blood

Oxygen concentration across the gill lamella of a bony fish

distance across gill lamella

Ventilation mechanism in bony fish

Bony fish have an internal bony skeleton and a flap covering the gills called the operculum.

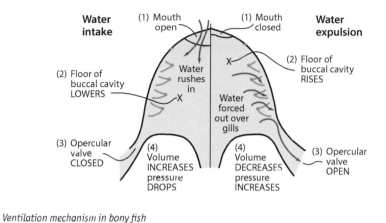

Ventilation mechanism in bony fish

Human respiratory system

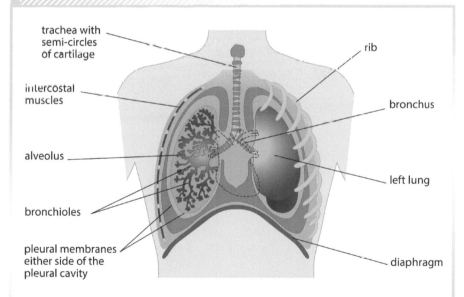

Human respiratory system

The trachea, which is supported by 20 incomplete cartilaginous rings, branches into two bronchi, each entering a lung. Bronchi branch into finer tubes called bronchioles, finally ending in alveoli where gas exchange takes place.

>> *Pointer*

Fish die out of water because the gills collapse and the filaments stick together greatly reducing the surface area for absorption of oxygen.

>> *Pointer*

Due to the efficiency of counter-current flow, more carbon dioxide diffuses from the blood into the water in bony fish than cartilaginous fish.

quickfire

③ Fill in the missing words.
Gas exchange in cartilaginous fish is .... effective than in bony fish. Water and blood flow in the .... direction, which is known as .... flow. Gas exchange occurs over .... of the gill lamellae so .... is reached, and .... oxygen is absorbed compared to counter-current flow. Bony fish can ....their gills by lowering and raising the .... of the mouth or buccal cavity. Cartilaginous fish can only ventilate their gills by continuously ....

Ventilation mechanism

Inspiration (active)

- External intercostal muscles contract moving ribs up and out, which pulls the outer pleural membrane outwards.
- Diaphragm contracts and flattens.
- This reduces the pressure in the pleural cavity and the inner pleural membrane moves outwards.
- This pulls on the surface of the lungs and causes the alveoli to expand.
- The alveolar pressure decreases to below atmospheric pressure, so air is drawn in.

Expiration (passive)

- External intercostal muscles relax so ribs move downwards and inwards, allowing the outer pleural membrane to move inwards.
- Diaphragm relaxes and moves upwards.
- This increases the pressure in the pleural cavity and the inner pleural membrane moves inwards.
- This pushes on the surface of the lungs and causes the alveoli to contract.
- The alveolar pressure increases to above atmospheric pressure, so air is forced out.

>> **Pointer**

The internal intercostal muscles are only used during forced expiration, e.g. blowing up a balloon, or during exercise.

Gas exchange in the alveolus

The alveoli are adapted for gas exchange by:

- very large surface area ~ 700 million alveoli
- very thin walls ~ 0.1μm
- surrounded by capillaries so short diffusion distance and good blood supply
- moist lining
- permeable to gases
- collagen and elastic fibres allow expansion and recoil.

A branch of the pulmonary artery brings deoxygenated blood to the alveoli and a branch of the pulmonary vein carries oxygenated blood from the alveoli back to the heart. The alveoli produce a surfactant, which lowers the surface tension preventing the alveoli from collapsing and sticking together, and allows gasses to dissolve.

>> **Pointer**

Surfactant is not produced by the foetus until around 23 weeks of pregnancy. A foetus born before this time will have difficulty in breathing.

>> **Pointer**

The gills are four times more efficient at extracting oxygen than the lungs. This is due to the fact that water contains much less dissolved oxygen in it than is found in air.

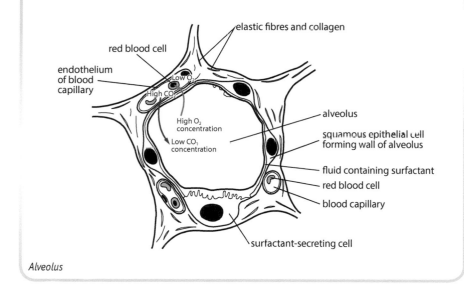

Alveolus

quickfire

④ Which blood vessel carries blood rich in carbon dioxide towards the alveoli?

Gas exchange in insects

Insects have a branched, chitin-lined system of tracheae with openings called spiracles. The chitin is arranged into rings, which allows the tracheae to expand and contract and act like bellows drawing air in and out of the insect's body. The spiracles, which are found in pairs on segments of the thorax and abdomen, can close during inactivity, and the presence of chitin, help to reduce water loss. Tracheole tubes come into direct contact with every tissue, supplying oxygen and removing carbon dioxide, so there is no need for haemoglobin. The ends of tubes are filled with fluid to allow gases to dissolve. Muscles in the thorax and abdomen contract/relax causing rhythmical movements that ventilate the tracheole tubes, maintaining a concentration gradient.

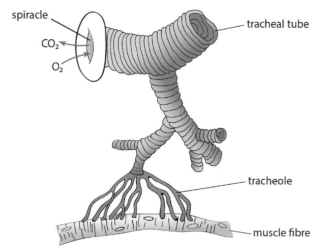

Tracheal system of an insect

>> *Pointer*

Insects have an exoskeleton of chitin covered in wax, this is impermeable to water and gases.

>> *Pointer*

The efficiency of the tracheal system in supplying oxygen to tissues limits insect size and shape as it relies upon diffusion, which is dependent upon diffusion distance.

⧀ *Grade boost*

Do not include a good blood supply as an adaptation of gas exchange surfaces in insects: they do not use it.

Gas exchange in plants

Plants require oxygen for respiration and carbon dioxide for photosynthesis: these gases are obtained by diffusion through the leaf. However, to reduce water loss, plants have a waxy cuticle that covers the surface of the leaf, which also prevents the diffusion of gases. Plants have pores called stomata found on the underside of most leaves that can open during the day to allow gas exchange, and close at night or during drought conditions to reduce water loss.

Key Term

Transpiration: the evaporation of water vapour from the leaves or other above-ground parts of the plant, out through stomata into the atmosphere.

Stomatal opening mechanism

The size of the pore (stoma) between the guard cells can be controlled to reduce water loss via **transpiration** by the guard cells that surround it.

1. Guard cells photosynthesise producing ATP.
2. Energy released from ATP is used to actively transport potassium ions *into* guard cells
3. This triggers starch (insoluble) to be converted to malate ions (soluble).
4. Water potential of guard cell is lowered so water enters cells by osmosis.
5. Guard cells expand and outer wall stretches more than the inner wall because it is thinner. This creates a pore between the two guard cells.
6. Reverse happens at night.

>> *Pointer*

Air spaces increase the rate of diffusion because it takes place in the gas phase.

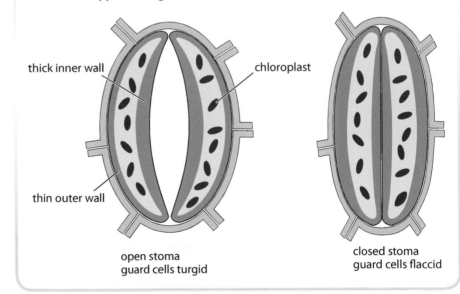

thick inner wall

chloroplast

thin outer wall

open stoma
guard cells turgid

closed stoma
guard cells flaccid

Adaptations of the leaf for gas exchange

- Leaves are thin and flat providing a large surface area to capture light and for gas exchange.
- Leaves have many pores called stomata (singular stoma) to allow exchange of gases.
- Spongy mesophyll cells are surrounded by air spaces that allow gases to diffuse.

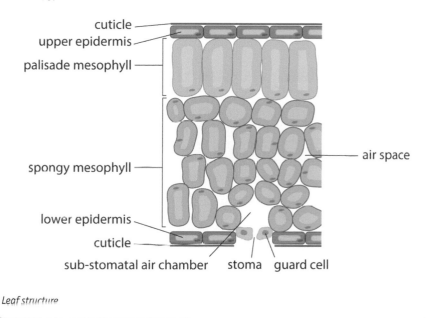

Leaf structure

If the capillary bringing deoxygenated blood to the alveolus has a partial pressure of carbon dioxide in the blood of 5.8 kPa, and the partial pressure of carbon dioxide in the blood in the capillary leaving the alveolus is 5.1 kPa, what must be the partial pressure of carbon dioxide in the alveolar air? Give a reason.

Grade boost

You could be asked to calculate the magnification or actual size from a drawing or microscope image:

$$\text{magnification} = \frac{\text{size of image}}{\text{actual size}}$$

$$\text{Actual size} = \frac{\text{size of image}}{\text{magnification}}.$$

quickfire

(5) True or false?

A. Surfactant raises surface tension.

B. Gills are less efficient at absorbing oxygen than the lungs.

C. When starch is converted to malate in guard cells, the water potential becomes more negative.

D. Spiracles can close to prevent water loss in plants.

E. Flatworms have a larger surface area than an earthworm with the same volume.

quickfire

(6) If a drawing of a leaf is 15 mm thick and shows a magnification of 50 times, what is the actual thickness of the leaf?

2.3a Adaptations for transport in animals

All transport systems in animals have a suitable medium to carry dissolved substances, which is aided by a pump to move the materials. Some systems (not insects) have a respiratory pigment, e.g. haemoglobin, to carry dissolved gases and use a system of vessels with valves to ensure a one-way flow to all parts of the body.

> **Key Term**
>
> **Haemocoel**: the main body cavity found in most invertebrates that contains a circulatory fluid.

Open circulatory systems

In open circulatory systems, blood does not move around the body in blood vessels. Instead, cells are bathed by blood or a fluid called haemolymph in a fluid filled space around the organs called a **haemocoel**, which returns slowly to the dorsal, tube shaped-heart, e.g. insects. There is no need for a respiratory pigment as oxygen is supplied directly to tissues via the tracheal system.

Open circulatory systems are relatively inefficient. In insects, they are not responsible for the distribution of respiratory gases.

Closed circulatory systems

Closed circulatory systems have the advantage of using vessels, so blood can be transported more quickly under a higher pressure, to all parts of the animal's body.

1. A single circulatory system involves blood passing through the heart once during its passage around the body. This type is found in fish where blood is pumped to the gills and onto the body organs before returning to the heart. A single circulatory system is also found in the earthworm where five pairs of 'pseudohearts' (which are thickened muscular blood vessels) pump blood from the dorsal vessel to the ventral vessel.

2. A double circulatory system involves blood passing through the heart twice: and in mammals involves one circuit which supplies the lungs where blood is oxygenated (pulmonary circulation), and a second circuit which supplies the body with oxygenated blood (systemic circulation). There are a number of advantages of a double circulation over a single circulation to meet the demands of mammals with a high metabolic rate: a higher blood pressure and faster circulation can be sustained in systemic circulation, and oxygenated and deoxygenated blood are kept separate, which improves oxygen distribution.

Grade boost

You should be able to compare the adaptations of the transport systems in earthworms, insects, fish and animals.

Structure and function of blood vessels

Blood vessel	Diagram
Arteries – carry blood AWAY from the heart. They have thick walls to resist the high blood pressure: elastic fibres stretch to allow the arteries to accommodate blood and the elastic recoil of the fibres pushes blood along the artery. The pressure in these arteries shows a rhythmical rise and fall, corresponding to ventricular **systole**. As the blood flows along the artery, friction with vessel walls causes the blood pressure and the rate of blood flow to decrease.	collagen fibres / thick layer of muscle and elastic fibres / endothelium / lumen (blood) / width 0.1-10mm
Arterioles – the main arteries continually branch to form smaller arteries and eventually arterioles. Arterioles have a large total surface area and relatively narrow lumen causing a further reduction in pressure and rate of blood flow. The important structure of an arteriole is the smooth muscle tissue, which can widen or narrow the lumen to increase or decrease blood flow.	
Capillaries – millions of capillaries form dense networks in tissues. They have a narrow lumen (8–10μm in diameter), but their total cross-sectional area is very large. As blood flows through the capillaries both blood pressure and the rate of blood flow decrease. This is due to the increase in total cross-sectional area and frictional resistance of blood flowing along the blood vessels. Their function is to supply oxygen and nutrients and absorb carbon dioxide and waste.	basement membrane / endothelium / lumen / red blood cell / width 8-10μm
Venules – are small veins that converge forming larger venules and eventually veins. They have a similar structure to veins and as they widen, the resistance to blood flow decreases allowing blood rate of flow to increase again.	
Veins – they carry blood *back* to the heart. The important structures of the vein are the semi-lunar valves, which prevent the backflow of blood ensuring that blood travels in one direction only. Although the pressure in veins is low, blood is returned to the heart due to the effects of surrounding skeletal muscle contracting, squeezing the vein, which reduces the volume, and increases the pressure inside the vein; this forces blood through the valve.	thin layer of smooth muscle and elastic fibres / collagen fibres / endothelium / valve / lumen / width 0.1-20mm

Grade boost

You need to relate a blood vessel's structure to its function.

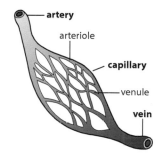

Blood vessels form a network that supplies oxygen to tissues.

Pressure changes in the different blood vessels can be seen in the graph:

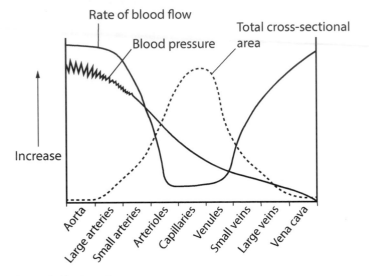

Pressure changes in blood vessels

The heart

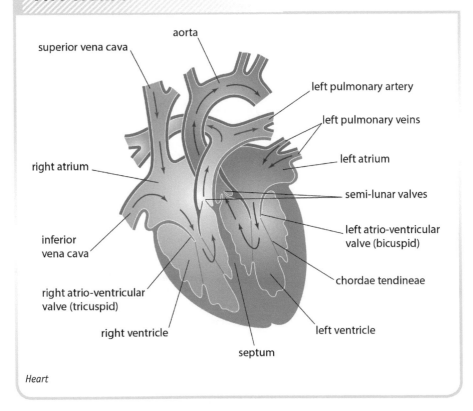

Heart

Blood flow through the heart

- Blood enters the heart from the head and body via the vena cava into the right atrium.
- The right atrium contracts (atrial **systole**) forcing blood through the right atrio-ventricular valve into the right ventricle, which is relaxed.
- The right ventricle contracts (ventricular systole) forcing blood out of the heart through the right semi-lunar valve to the lungs via the pulmonary artery.
- Oxygenated blood returns from the lungs to the heart via the pulmonary vein and enters the left atrium when the left atrium is relaxed (total **diastole**).
- The left atrium contacts, forcing blood through the left atrio-ventricular valve into the left ventricle, which is relaxed.
- The left ventricle contracts, forcing blood out through the left semi-lunar valve into the aorta and then to the rest of the body.
- This describes one circuit that blood takes. During the cardiac cycle, both atria contract simultaneously followed by the contraction of both ventricles.
- Valves ensure that blood flows in a unidirectional manner, i.e. they prevent backflow of blood.

Cardiac cycle

- Left atrium contracts so volume of atrium decreases and pressure increases.
- When blood pressure in left atrium exceeds that in left ventricle, blood flows into the left ventricle.
- The ventricle then contracts (ventricular systole) and pressure rises in left ventricle as volume decreases.

Cardiac cycle

- As the ventricle contracts blood is pushed against the atrioventricular valves closing them and preventing blood flow back to the atria (see 1 on diagram).
- When pressure in left ventricle exceeds that in the aorta, the left semi-lunar valve opens (see 2 on diagram) and blood flows out into the aorta.
- Left ventricle then relaxes (diastole) so its volume increases and pressure falls.
- When pressure in ventricle drops below that of the aorta, blood tries to flow back into the ventricle from the aorta, pushing against the left semi-lunar valve and closing it (see 3 on diagram).
- When pressure in left ventricle drops below that in the left atrium, the left atrio-ventricular valve opens (see 4 on diagram) and the cycle begins again.
- Remember: blood always flows from a region of *high* pressure to *low* pressure *unless* a valve prevents it.

>> **Pointer**

A good way to remember that **di**astole means relaxation is that when you **die**, you are very relaxed!

>> **Pointer**

Cardiac output: this is the stroke volume × heart rate – in other words, the total volume of blood pumped by the heart per minute is the volume of blood pumped per beat multiplied by the number of times the heart beats in a minute.

 Grade boost

You need to be able to describe these changes and identify where the different valves in the heart open and close.

Grade boost

Don't confuse cardiac cycle with control of heartbeat.

Bundle of His: modified cardiac muscle fibre passing from the AVN to the base of the ventricle through the septum of the heart.

Purkinje fibres: network of fibres in the wall of the ventricles.

>> *Pointer*

You can calculate the heart rate by measuring the time taken from one point on the ECG trace to the next, e.g. R to R (in this case 0.30s to 1.15s = 0.85s).

$$\text{Rate} = \frac{60}{0.85} = 71\text{bpm}$$

quickfire

② Match the following labels of an ECG trace: T, P, QRS, TP, to the following four statements:
1. Voltage change associated with contraction of atria.
2. Contraction of ventricles.
3. Repolarisation of ventricle muscles.
4. The filling time.

>> *Pointer*

The significance of the delay created by the wave of excitation arriving at the AVN, and it spreading via the Bundle of His is that the atria are fully emptied *before* the ventricles contract *and* to ensure that the ventricles contract from the base upwards.

Control of heartbeat

The heart is myogenic – the heartbeat is initiated within the cardiac muscle itself and is not dependent upon external stimulation. It can, however, be regulated, and this is accomplished by the sinoatrial node which initiates a wave of excitation across both atria.

Stage of cardiac cycle	Details
Atrial systole	<ul><li>Wave of excitation spreads out from the sinoatrial node (SAN) across both atria</li><li>Both atria start contracting</li><li>Wave cannot spread to ventricles due to layer of connective tissue</li><li>Wave spreads via the atrioventricular node (AVN), through the **Bundle of His** to apex of ventricle</li></ul>
Ventricular systole	<ul><li>The Bundle of His branches into **Purkinje fibres** carrying wave upwards through ventricle muscle causing it to contract</li><li>Ventricle contraction is therefore delayed and contraction is from base upwards</li></ul>

These events can be seen in an ECG (electrocardiogram).

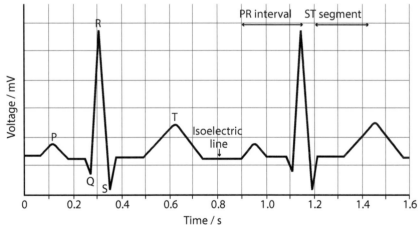

ECG

P = voltage change associated with contraction of atria

QRS = depolarisation and contraction of ventricles

T = repolarisation of ventricle muscles

The filling time is represented by the gap between T and the next P wave (labelled the isoelectric line)

Changes in this trace can be used to diagnose problems with the heart, e.g. a patient suffering a heart attack (myocardial infarction) may show depression in the S–T segment.

Blood

Blood consists of plasma (55%) and cells (45%). By far the majority of the cells present are red blood cells (erythrocytes), which contain haemoglobin. The remainder (<5%) are white cells and platelets. In humans, the red blood cells have a biconcave shape, which increases their surface area for the absorption and release of oxygen, and do not possess a nucleus. This means that they can carry more haemoglobin, but limits their life, so they have to be continually produced. There are two main types of white blood cells: granulocytes, which are phagocytic, and lymphocytes that develop into cells that produce antibodies.

Plasma is 90% water, and contains dissolved solutes, e.g. glucose and amino acids, hormones and plasma proteins. It is responsible for the distribution of heat and transport of carbon dioxide as HCO_3^- ions. Excretory products such as urea are also transported dissolved in the plasma.

Erythrocytes

Transport of oxygen

Haemoglobin binds to oxygen in the lungs, and releases it to the respiring tissues. This reversible reaction can be shown by:

oxygen + haemoglobin ⟷ oxyhaemoglobin

The association and dissociation of oxygen with haemoglobin is influenced by a number of factors and is shown by a dissociation curve.

Each molecule can accommodate four molecules of oxygen ($4O_2$), one attached to each of the four haem groups. As oxygen molecules bind, the haemoglobin molecule changes slightly, making it easier for the next one to bind. This is known as cooperative binding, and can be seen by the steep part of the curve. The fourth and final oxygen molecule is more difficult to bind, and requires a large increase in **partial pressure** of oxygen to accomplish this: this is shown by the plateau on the graph, giving the curve a sigmoid (S) shape.

> **》 Pointer**
>
> It may be helpful to think of partial pressure like concentration. As the partial pressure increases, more is present.

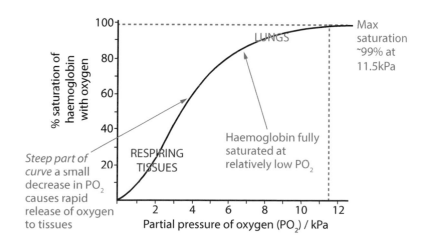

Dissociation curve

Key Term

Affinity: the degree to which two molecules are attracted to each other.

Grade boost

When answering questions on oxygen dissociation, remember to talk in terms of haemoglobin's affinity for oxygen and the consequences of this.

≫ Pointer

Using a pencil draw lines from the same partial pressure of oxygen up to the curve so you can compare the percentage saturation.

≫ Pointer

Low oxygen environments (llama/lugworm) curve is to the *left*.

≫ Pointer

Myoglobin is an oxygen store found in muscles. It has a very high affinity for oxygen and only releases oxygen at very low partial pressures. It is found even further to the left of a llama's haemoglobin.

③ What is the significance of the haemoglobin curve shifting to the left compared to a shift to the right?

④ Suggest one change that could be seen in the blood of an athlete training at high altitude.

The graph shows how haemoglobin's **affinity** for oxygen changes with partial pressure: at high partial pressures, the affinity is high, so oxyhaemoglobin does not easily release its oxygen. At low partial pressures, oxygen is released rapidly to the respiring tissues where it is needed, because haemoglobin's affinity for oxygen is low.

Low oxygen environments

In low oxygen environments, such as high altitudes or muddy burrows, animals have become adapted by evolving haemoglobin with a higher affinity for oxygen than normal haemoglobin. Animals such as the llama and lugworm have oxygen dissociation curves to the *left* of normal (see graph) meaning that their haemoglobin is more saturated at the same partial pressure of oxygen than normal, i.e. it is more able to pick up oxygen.

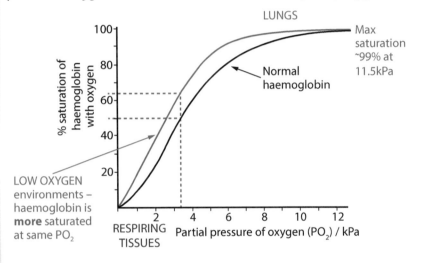

Low oxygen environments

Foetal haemoglobin also has a higher affinity for oxygen than normal (maternal blood) and so is also to the left. This means that it is able to absorb oxygen from the mother's blood via the placenta.

Effects of carbon dioxide on dissociation

When carbon dioxide concentrations in the blood rise during exercise, the haemoglobin dissociation curve shifts to the right (see graph). This is because the haemoglobin's affinity for oxygen is reduced, so more oxygen is released at the same partial pressure of oxygen. This supplies oxygen more quickly to respiring tissues, where it is needed. This is called the **Bohr effect**, and is easily explained when you understand how much of the carbon dioxide is carried in the blood.

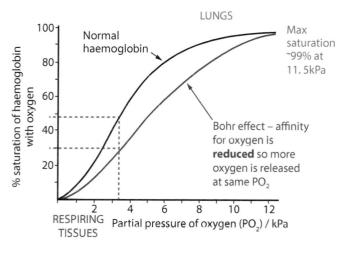

Bohr effect graph

Carbon dioxide is carried in three ways:

- dissolved in plasma (5%)
- as HCO_3^- ions (85%) in the plasma
- bound to haemoglobin as carbamino-haemoglobin (10%).

The majority of carbon dioxide is carried as HCO_3^- ions, which is formed via a series of reactions within the red blood cell itself.

Reactions in a red blood cell

1. Carbon dioxide diffuses into the red blood cell.

2. Carbonic anhydrase catalyses the reaction between carbon dioxide and water forming carbonic acid.

3. Carbonic acid dissociates into HCO_3^- and H^+ ions.

4. HCO_3^- diffuses out of the red blood cell.

5. Cl^- ions diffuse into the cell to maintain the electrochemical neutrality: this is called the chloride shift.

6. H^+ ions combine with oxyhaemoglobin forming haemoglobinic acid (HHb) and releasing oxygen.

7. Oxygen diffuses out of the cell.

extra

Each red blood cell contains around 280×10^6 haemoglobin molecules when fully saturated at 95% saturation. Each haemoglobin molecule can carry four oxygen molecules. How many oxygen molecules are released from one red blood cell if the percentage saturation of haemoglobin is 44%?

Key Term

Bohr effect: the movement of the oxygen dissociation curve to the right as a result of higher partial pressure of carbon dioxide. Haemoglobin shows a reduced affinity for oxygen.

≫ Pointer

More CO_2 in the blood = more HCO_3^- and H^+ ions produced inside the red blood cell = more HHb produced and so more O_2 released.

quickfire

⑤ What is the significance of the chloride shift?

Formation of tissue fluid

Tissue fluid is formed from the plasma that is found in the blood. It contains water, salts, glucose, amino acids and dissolved oxygen and bathes the cells in the capillary bed. It is formed by the following processes:

1. Hydrostatic pressure created by blood pressure at the arteriole end forces these materials out of the capillaries through pores in their walls. Plasma proteins are too large to leave. The water potential of the blood is lower than that of the tissue fluid, tending to draw water into the capillary and this decreases along the capillary as water leaves, but this is less than the hydrostatic pressure so net movement is *out* of the capillaries.

2. Most of the water is reabsorbed by osmosis at the venule end of the capillary bed. Here osmotic pressure (created by water potential gradient) exceeds hydrostatic pressure in capillaries so net movement of fluid is *into* the capillary. Carbon dioxide is reabsorbed by diffusion.

3. Excess tissue fluid drains into the lymphatic system and returns to the venous system via the thoracic duct, which empties into the left subclavian vein in the neck.

>> **Pointer**

Malnourished people have lower concentrations of plasma proteins and so reabsorb *less* water by osmosis at the venule end of the capillary because the osmotic pressure is lower. This leads to oedema (fluid retention) and a condition called Kwashiorkor.

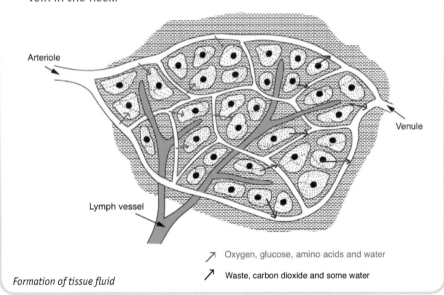

Formation of tissue fluid

↗ Oxygen, glucose, amino acids and water

↗ Waste, carbon dioxide and some water

2.3b Adaptations for transport in plants

Vascular tissue in plants is made up of two main types of tissue: xylem and phloem. Xylem is responsible for the transport of water and mineral ions as well as providing support, whilst phloem is responsible for translocation of organic solutes, e.g. sucrose, and amino acids. Their arrangement is organised differently in roots, stems and leaves.

Part of plant	Arrangement of vascular tissue	Diagram
Root	Xylem is arranged centrally into a star shape with phloem outside it. This helps to anchor plant into the soil, resisting pulling forces.	
Stem	Arranged towards the periphery in a ring, which provides support to resist bending.	
Leaf	Arranged in the midrib giving both resistance to tearing and flexibility.	

quickfire

① State two functions of xylem.

⚜ Grade boost

Be able to recognise the xylem and phloem from photomicrographs and electron micrographs of roots, stems and leaves.

Structure of xylem

Water is conducted through vessels and tracheids, which are dead cells due to lignin deposition in the walls. Fibres provide support, and xylem parenchyma acts a packing tissue. Tracheids are present in flowering plants (**angiosperms**), ferns and conifers, but vessels are only present in flowering plants.

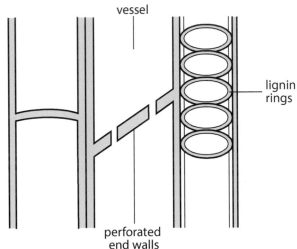

Structure of xylem

Water uptake by the roots

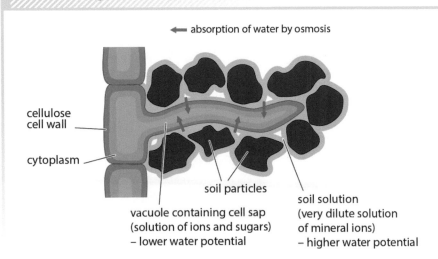

Absorption of water by root hair cell

Root hair cells are adapted for the uptake of water by having a large surface area. Water enters the root hair cells by osmosis, because the soil solution has a higher water potential than the vacuole of the hair cell which contains ions and sugars.

Water moves across the cortex of the root from the epidermis towards the xylem in the centre via three different pathways:

1. Apoplast pathway – the most significant route, involves water moving between the spaces in the cellulose cell wall.

2. Symplast pathway – water moves through the cytoplasm and plasmodesmata (strands of cytoplasm through cell wall pits).

3. Vacuolar pathway – is a minor route and involves water passing from vacuole to vacuole.

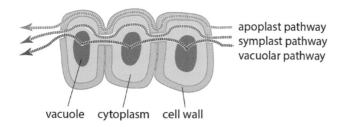

Three pathways that water takes across the root cortex in plants

The presence of lignin in the cell walls of the xylem vessels waterproofs them. It will also prevent water from entering the xylem via the apoplast pathway. In the root, the pericycle is surrounded by a single layer of cells called the endodermis, which forms a ring surrounding the vascular tissue in the centre of the root. The cell walls of the endodermis are impregnated with suberin, forming an impermeable band known as the Casparian strip that drives water from the apoplast pathway into the cytoplasm. The endodermis helps to regulate the movement of water, ions and hormones into and out of the xylem.

The water potential of endodermal cells is raised by water being forced into them by the Casparian strip and the active transport of sodium ions into the xylem. This lowers the water potential of fluid in the xylem, forcing water into the xylem by osmosis: this is known as root pressure.

Uptake of minerals

Minerals including nitrates and phosphates are actively transported into the root hair cells against their concentration gradient. They can also pass along the apoplast pathway in solution. Once they reach the Casparian strip they enter the cytoplasm via active transport and then pass via diffusion or active transport into the xylem.

> **Grade boost**
> It is important to know whether processes involving water transport are active or passive.

> ## quickfire
> ② Explain why water passing along the apoplast route is diverted when it reaches the endodermis.

> ## quickfire
> ③ What would the effect be on mineral uptake if a respiratory inhibitor such as cyanide was used?

Transpiration: the evaporation of water vapour from the leaves or other above-ground parts of the plant, out through stomata into the atmosphere.

Adhesive forces: are created between the charges on the water molecules and their attraction with the hydrophilic lining of the vessels.

Cohesive forces: are created by the attractive forces between water molecules due to their dipolar charges forming hydrogen bonds.

Capillarity: the movement of water up narrow tubes by capillary action.

Movement of water from roots to leaves

The tallest trees are over 100m tall, and therefore it is quite a feat to transport hundreds of litres of water that distance up to the leaves against gravity every day. The cohesion-tension theory explains how water moves up the xylem. The main mechanism that pulls water up the stem is **transpiration**, which is a passive process. Transpiration pull relies on: **adhesive forces** between water molecules and xylem, and **cohesive forces** between water molecules, root pressure and **capillarity** are also involved but alone would not be sufficient to raise water up the xylem to any significant height.

Transpiration pull is created as water evaporates from the leaf air space through the stomata (although there is some diffusion through the cuticle); water is drawn from inside the cells lining the space by osmosis. These cells now have a lower water potential and so draw water from adjacent cells by osmosis, and this continues across the leaf until water is drawn from the adjacent xylem vessel. As water is drawn out of the xylem, water molecules are 'pulled' up to replace those lost due to cohesive forces that exist between water molecules. Water molecules enter the xylem to replace those moving up by osmosis from the endodermal cells, and water crosses the cortex from the root hair by the same method as in the leaf cells.

quickfire

④ List the main forces that enable transpiration pull to occur.

quickfire

⑤ For each of the following conditions, state whether the rate of transpiration increases or decreases, and explain why.
   A. Wind speed drops.
   B. It starts to rain.
   C. The temperature rises.

Factors affecting the rate of transpiration

Plants have to balance water loss by transpiration with the need to get water and mineral ions to the leaves themselves. Therefore, water loss is inevitable. Four factors affect the rate at which water is lost by transpiration:

Factor	Effect
Temperature	Increasing temperature causes the water molecules to gain more kinetic energy therefore increasing the rate of diffusion out into the atmosphere through the stomata.
Humidity	As the humidity of the air outside the leaf increases, the difference between the inside and outside of the leaf decreases, reducing the diffusion gradient.
Air movement	As air speed increases, saturated air is removed from the leaf surface more quickly, therefore increasing the diffusion gradient.
Light Intensity	Increasing light intensity will increase stomatal opening. See page 84 (stomatal opening mechanism).

As a result, plants exposed to a bright, hot, dry, windy day will show the highest rates of transpiration.

Practical – comparing rates of transpiration using a potometer

A potometer actually measures the rate of water uptake. Some water will be used in photosynthesis, but if cells are turgid, this uptake rate is approximated as the transpiration rate. When setting up a potometer it is important to:

- Cut the stem and fit it to the potometer underwater, as this prevents the formation of any air bubbles in the xylem vessels.
- Seal all joints with Vaseline to prevent air entry.
- Blot leaves dry as any water on the leaf surface could create a humid layer.

Then introduce an air bubble at the end of the capillary tube, and measure the distance it travels in a set period, e.g. one minute. The volume can be calculated easily if the diameter of the capillary tube is known. As with any experiment, you should carry out repeats.

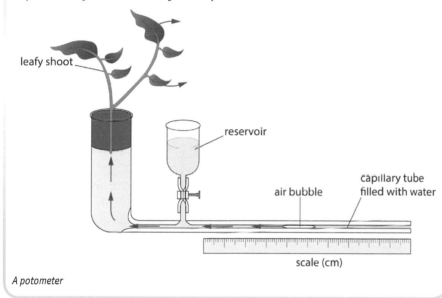

leafy shoot

reservoir

air bubble

capillary tube filled with water

scale (cm)

A potometer

 Pointer

To calculate volume of water taken in by a plant using a potometer use the equation **volume = πr²d** (where π = 3.14, r = radius of capillary tube, and d = distance bubble travels). Remember:

$$\frac{diameter}{2} = radius$$

 Grade boost

You need to know how to set up a potometer including precautions taken, and how to collect valid results.

Plant adaptations to living with varying water availability

Mesophytes

Mesophytes live in temperate regions with an adequate water supply but must survive times of the year when water is scarce or unavailable, e.g. water is frozen. They conserve water by:

- Closing stomata if water is scarce, as they cannot maintain turgor in guard cells.
- Shedding leaves and becoming dormant during winter.
- Overwintering beneath the ground as bulbs or corms.
- Annual plants producing seeds that can overwinter.

Xerophytes

Xerophytes, e.g. Marram grass, are plants that have become adapted to living in dry environments by reducing water loss. Adaptations include:

- Sunken stomata, which trap humid air, reducing the water potential gradient between air spaces inside the leaf and the outside air.
- Hairs around stomata, which trap water vapour, reducing the water potential gradient between the leaf and the air.
- Rolled leaves, which reduces the surface area over which transpiration occurs. Some plants take this to extremes by reducing leaves to spines and using the stem to photosynthesise, e.g. cacti.
- Thick cuticle, which further reduces water loss from leaf surface.

Hydrophytes

Hydrophytes grow partially or fully submerged in water, so lack of water is never a problem, but ensuring that they receive adequate light and carbon dioxide for photosynthesis is. The water lily is adapted by:

- Having stomata on the upper leaf surface, which is in contact with the air.
- Stems and leaves have large air spaces providing buoyancy and a reservoir of oxygen and carbon dioxide.
- Having poorly developed xylem tissue as there is no need to transport water as it is all around.
- Leaves have little or no cuticle as water loss is not a problem.
- Support tissue is not needed as water is a supportive medium.

quickfire

⑥ Explain how sunken stomata reduce transpiration.

quickfire

⑦ Describe how insects and plants are able to reduce water loss in similar ways.

Translocation

The products of photosynthesis are transported in the phloem as sucrose from where they are produced (the source) to where they are used or stored as insoluble food reserves e.g. starch (the sink). The phloem also transports amino acids.

Structure of phloem

Phloem is a living tissue and consists of three main types of cells:

Sieve tubes – walls perforated with pores to produce longitudinal tubes that contain cytoplasm but no nucleus, and most organelles disintegrate during their development. The end walls do not break down but instead become perforated by pores, forming the end plates.

Companion cells – dense cytoplasm with nucleus and many mitochondria and are connected to each sieve tube by plasmodesmata.

Phloem parenchyma acting as a packaging tissue.

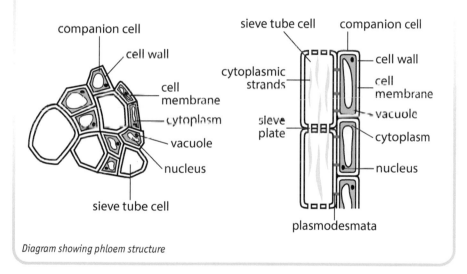

Diagram showing phloem structure

quickpire

⑧ Suggest why companion cells contain many mitochondria.

Evidence that phloem is the vessel involved in translocation

A number of experiments have been used to show which vessels transport solutes. By radioactively labelling carbon dioxide using ^{14}C, products and their paths can be traced by exposing the plant to X-ray film. These are called autoradiographs. Ringing experiments have also been used whereby the outer ring of stem is cut to remove the phloem whilst leaving the xylem behind. A bulge forms above the ring suggesting that sugar moves down the stem in the phloem. In early experiments, aphids were allowed to feed on plants, and then were anaesthetised before removing the head leaving the feeding stylet in place. Analysis of the liquid extruding from the stylet showed that it was sucrose.

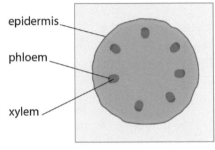

epidermis

phloem

xylem

section of stem
placed against photographic
film in the dark

developed film emulsion
is fogged by the presence
of radioactivity in the phloem

Results from ^{14}C experiment

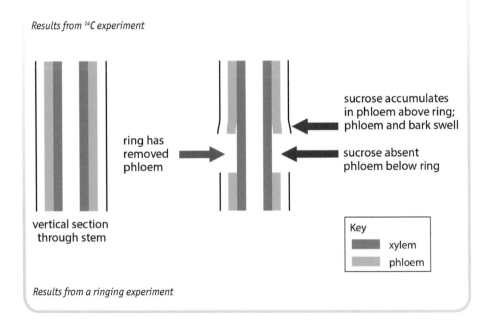

ring has
removed
phloem

sucrose accumulates
in phloem above ring;
phloem and bark swell

sucrose absent
phloem below ring

vertical section
through stem

Key
xylem
phloem

Results from a ringing experiment

quickfire

⑨ Name two cell types present in phloem tissue.

Mass flow theory

A number of theories have been put forward to explain how organic solutes are transported. The mass flow hypothesis is the most accepted theory put forward, but still fails to explain how sucrose and amino acids are transported at different rates in opposite directions in the same phloem vessel, or how transport occurs thousands of times faster than is possible by diffusion.

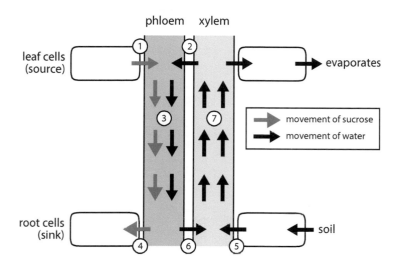

Mass flow theory

1. Photosynthesising cells (source cells) produce glucose, which is converted into sucrose, which lowers the water potential of the cell. As water enters the cell by osmosis, hydrostatic pressure forces sucrose into the phloem sieve tube.

2. By increasing the level of solutes in the phloem, the water potential (Ψ) is lowered and water moves in from the adjacent cells and xylem, by osmosis, down a water potential gradient. This raises the *hydrostatic pressure* in the phloem so that it has a higher pressure.

3. Sucrose and dissolved solutes move by mass flow from a high to a low hydrostatic pressure, down a pressure gradient.

4. At the roots/growing points (sink cells) the sucrose diffuses into the cells down a concentration gradient, so is therefore removed from sieve tubes. In the sieve tubes, it is converted to starch for storage or converted to glucose to be respired. The loss of sucrose from phloem raises the water potential higher than in the xylem and adjacent cells.

5. Water enters the xylem by osmosis.

6. Water also moves from phloem to xylem down a water potential gradient, causing a reduction in the hydrostatic pressure.

7. Water moves up the xylem by transpiration.

Other theories involving the use of protein filaments and cytoplasmic streaming, which could account for bidirectional transport, have been suggested.

quickpire

10 Where would you expect to find sink cells in a plant?

Grade boost
You need to be able to describe some aspects of translocation which the mass flow hypothesis cannot account for.

Grade boost
You should be able to interpret and explain results obtained from investigations into phloem transport.

2.4 Adaptations for nutrition

Modes of nutrition

Mode of nutrition	Organism	How they obtain energy
Autotrophic – make own organic food from simple inorganic raw materials	Photoautotrophic	Use light energy to perform photosynthesis. Examples include green plants, Protoctista and some bacteria.
	Chemoautotrophic	Use energy from chemical reactions. Examples include some prokaryotes.
Heterotrophic – consume complex organic molecules produced by autotrophs	Saprotrophic	Feed on dead or decaying matter by secreting enzymes extracellularly and then absorbing the products, e.g. *Rhizopus* (bread mould).
	Parasitic	Obtain nutrition from another living organism called the host over a long period of time, whilst causing it harm. Endoparasites live within the host's body, e.g. tapeworm, *Taenia*. Ectoparasites live on the surface, e.g. human head louse, *Pediculus*.
	Holozoic	Form of nutrition used by most animals where they ingest and then digest food, absorbing nutrients. They possess a specialised digestive system. Examples include herbivores (plant material), carnivores (animal material), omnivores (plant and animal material) and detritivores (dead or decaying material).

Nutrition in unicellular organisms

Protoctista such as amoeba are holozoic heterotrophs. They absorb nutrients directly through their cell membrane by diffusion, ingesting larger molecules by endocytosis and fluids by pinocytosis into food vacuoles. Lysosomes fuse with the vacuoles and release digestive enzymes. Nutrients are absorbed through the membrane of the food vacuole and waste is egested by exocytosis.

Nutrition in multicellular organisms

Some larger organisms have a single body opening; e.g. *Hydra*, which lives in fresh water. Tentacles paralyse prey, e.g. *Daphnia*, and move it into the hollow body cavity through the mouth. Protease and lipase enzymes digest the food extracellularly, and the products are absorbed before the indigestible remains are egested back out through the mouth.

More developed organisms possess a tube gut, ingesting at one end, egesting at the other, with the most advanced possessing specialised regions.

Human digestive system

The gut consists of a long hollow muscular tube, through which food passes along by **peristalsis**. Different regions of the gut are specialised to perform four main functions:

1. Ingestion – taking food into the body via the mouth bringing it into contact with the digestive surface.

2. Digestion – results in large biological molecules being **hydrolysed** to smaller molecules that can be absorbed across cell membranes. Digestion begins with mechanical digestion in the mouth involving the teeth, which breaks large food pieces into smaller pieces. Digestion is then completed by enzymes.

3. Absorption – passage of nutrient molecules through the wall of the gut into the blood.

4. Egestion – elimination of undigested material, e.g. cellulose fibre.

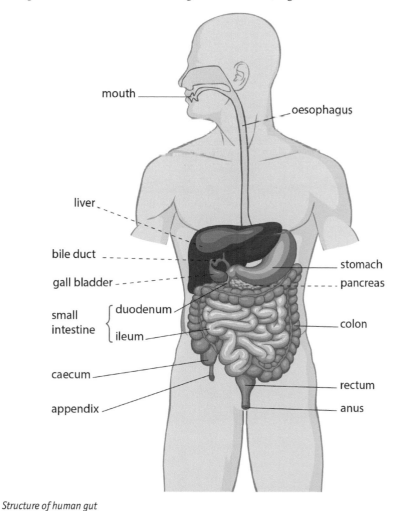

Structure of human gut

Key Terms

Peristalsis: the wave of coordinated contraction and relaxation of smooth muscles in the gut.

Hydrolysis: reaction involving the chemical addition of water to break the bond formed during condensation.

Grade boost

Mechanical digestion in the mouth mixes salivary amylase with food, and the teeth break large pieces of food into smaller pieces NOT smaller molecules.

Grade boost

Egestion is getting rid of undigested food via faeces. Excretion is eliminating waste made within the body, e.g. urea, and carbon dioxide.

Structure of gut wall

The human gut wall consists of four basic layers, which vary in proportion in different specialised regions.

Layer	Function
Serosa	Outermost layer consisting of tough connective tissue, which protects the gut and reduces friction from other abdominal organs.
Muscle	Consists of two layers: circular and longitudinal smooth muscles which contract in a coordinated fashion to push food along by peristalsis.
Submucosa	Connective tissue containing blood and lymph vessels to take away the absorbed products of digestion. Nerves are also present, which co-ordinate muscular contractions.
Mucosa	Innermost layer lining the gut; it secretes mucus (lubrication and protection from enzymes). Depending upon the region, it secretes enzymes and absorbs digested food and nutrients.

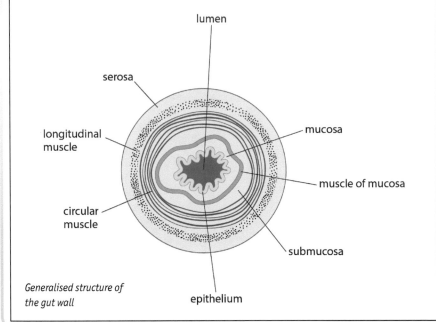

Generalised structure of the gut wall

① Name four layers of the gut wall, in order from the lumen outwards.

⬆ *Grade boost*

You must be able to label a diagram showing a section through the wall of the gut.

Digestion

The digestion of different food groups and their subsequent absorption takes place in different parts of the gut. This is looked at in more detail on page 108. Carbohydrate digestion begins with the hydrolysis of starch in the mouth by salivary amylase, whilst protein digestion begins in the stomach by the action of pepsin. Digestion of the different food groups and the enzymes involved are shown in the table below.

Food	Process
Carbohydrates	Amylase hydrolyses starch to maltose, and then maltase hydrolyses maltose to glucose. Sucrase hydrolyses sucrose to glucose and fructose. Lactase hydrolyses lactose to glucose and galactose. Starch —Amylase→ Maltose —Maltase→ Glucose Polysaccharide · Disaccharide · Monosaccharide
Proteins	Proteins are digested into polypeptides, dipeptides and eventually amino acids. The enzymes involved are peptidases, named according to where they break peptide bonds. Protein —Endopeptidase→ Polypeptide —Exopeptidase→ Dipeptide Dipeptide —Dipeptidase→ Amino acids
Fats	Fats are emulsified by bile and then hydrolysed to fatty acids and glycerol. Lipids (large globules) —Bile→ Lipids (small globules) —Lipase→ Fatty acids + glycerol

Grade boost

Amylase is breaking the glycosidic bonds within amylose and amylopectin, hence its name.

Grade boost

Bile is NOT an enzyme.

Pointer

Endo means internal, exo means external. This will help you to remember what part of the protein molecule is hydrolysed by endopeptidases and exopeptidases.

Regional specialisation of the mammalian gut

1. The mouth (buccal cavity)

   This is where digestion begins. Teeth mechanically digest food, the tongue mixes this with saliva and rolls it into a bolus for swallowing. The saliva contains the enzyme, amylase, and mucus, which lubricates the food. The enzyme amylase initiates starch digestion.

2. The oesophagus muscles contract to move the food towards the stomach via peristalsis.

3. In the stomach, food is digested for about four hours by muscular action from the stomach walls and gastric juice, which contains hydrochloric acid (from oxyntic cells in gastric pits) and pepsin. Pepsin is an endopeptidase that is secreted in an inactive form pepsinogen, and activated by H^+ ions: this prevents pre-digestion. The acidic pH of around 2 also kills bacteria. Mucus is produced by goblet cells in the gastric pits, which lubricates food and protects the lining.

4. The duodenum is the first part of the small intestine, receiving secretions from the liver and pancreas. Bile, which contains bile salts, neutralises acidic food from the stomach and emulsifies fats. Pancreatic juice is slightly alkaline (due to the presence of sodium hydrogen carbonate) and is secreted by islet cells in the pancreas, entering the duodenum via the pancreatic duct. It contains endopeptidases and trypsinogen (which is inactive, and converted to the active form trypsin by enterokinase), amylase and lipase. Brunner's glands at the base of the crypts of Lieberkühn produce alkaline secretions that also neutralise acidic food from the stomach. The mucosa of the small intestine is heavily folded to form villi. In the duodenum, endopeptidases and exopeptidases are secreted by cells at the tips of the villi, and peptidases bound to epithelial cells complete the digestion to amino acids. Maltase, lactase and sucrase enzymes are also bound to the epithelial cells and complete the digestion of carbohydrates.

Key Term

Emulsification: large fat droplets are broken into smaller droplets which increases the surface area for lipase action.

≫ Pointer

Co-transport involves transporting two different molecules together, e.g. glucose and sodium ions, and is the mechanism by which glucose is absorbed in the ileum of mammals – see page 35.

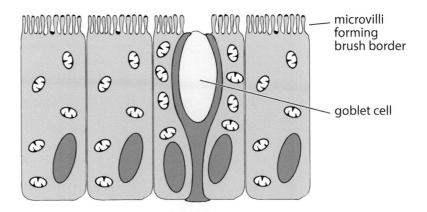

microvilli forming brush border

goblet cell

Structure of epithelial cells lining the small intestine

5. The ileum is the second part of the small intestine, which is responsible for the absorption of digested food. Villi and microvilli greatly increase the surface area of the ileum (over 600 times) for absorption by diffusion, facilitated diffusion, co-transport and active transport and for the action of membrane bound enzymes:

   • Glucose enters into epithelial cells by co-transport and active transport and by facilitated diffusion into capillary of villus.

   • Amino acids enter by active transport into epithelial cells and then by facilitated diffusion into the capillary of villus.

   • Fatty acids and glycerol enter epithelial cells via diffusion where they recombine into triglycerides and enter lacteal of villus. In these epithelial cells, smooth endoplasmic reticulum is highly developed to assist in this process.

   • Water is absorbed by osmosis into epithelial cells and the capillary of villus.

   • Water-soluble vitamins (e.g. B and C) are absorbed directly into the blood, whilst fat-soluble vitamins (e.g. A, D and E) are absorbed into the lacteals by diffusion.

quicKpire

② Into which parts of the villus are the following products absorbed?
   A. glucose
   B. fatty acids and glycerol
   C. amino acids.

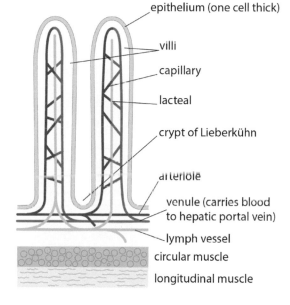

epithelium (one cell thick)

villi

capillary

lacteal

crypt of Lieberkühn

arteriole

venule (carries blood to hepatic portal vein)

lymph vessel

circular muscle

longitudinal muscle

LS ileum wall

6. The large intestine (appendix, caecum, colon, rectum and anus) has small villi present and is responsible for absorption of water and the formation of faeces, which is stored in the rectum until it is egested. Mutualistic bacteria present in the colon are responsible for the production of vitamin K and folic acid. Glucose and amino acids are transported by the hepatic portal vein to the liver where they are processed. The lacteals drain into the lymphatic system which drains into the blood via the thoracic duct in the right subclavian vein.

Complete the following table which summarises digestion of the different food groups in humans.

Food	Region of gut	Enzyme(s)	Site of production	pH	Substrate	Products	How Absorbed
Carbohydrate	Mouth	Amylase	....	7	....	Maltose	
	.... (1st part of small intestine)	Amylase	....	7	....	Maltose	
	.... (2nd part of small intestine)	....	.... mucosa	8.5	Maltose Sucrose Lactose	.... + ... + ...	Glucose enters by .... into .... cells and by .... into capillary of villus
Protein	Stomach	....	Gastric glands	2	....	Polypeptides	
	.... (1st part of small intestine)	endopeptidases	....	7	Polypeptides	Polypeptides	
	.... (2nd part of small intestine)	endopeptidases & exopeptidases	.... mucosa	8.5	....	....	Amino acids enter by active transport into .... cells and then by .... into the capillary of villus

Lipid*	.... (1st part of small intestine)	....	7	Lipids	.... &	Fatty acids and glycerol enter .... cells via diffusion. They recombine into .... and enter .... lacteal of villus
	.... (2nd part of small intestine)	.... mucosa	8.5	Lipids	.... &	
....	-NA-	-	-	-	-	Provides bulk and stimulates ....
Water	-NA-	-	-	-	-	By .... into villi and colon

NB – bile emulsifies fats producing smaller droplets, which increases the surface area for lipase action.

Mucus lubricates food and protects gut wall from enzymes/acid (stomach).

Adaptations to different diets

The dentition and gut structure of different animals is adapted to reflect their diet. Animals that eat only meat have different dentition and gut structure from animals that eat a diet consisting purely of vegetation.

Carnivores

Carnivores have evolved sharp incisors to tear flesh, pointed canines to pierce flesh and kill prey, and specialised molars (3rd premolar upper jaw and 1st molar lower jaw) called **carnassials** that shear flesh and bone. They have powerful jaw muscles, which move the lower jaw vertically up and down, and are able to open their jaws wide to accommodate large prey animals. Their intestines are relatively short, because the main constituent in their diet is protein, which is relatively easily digested.

Carnivore

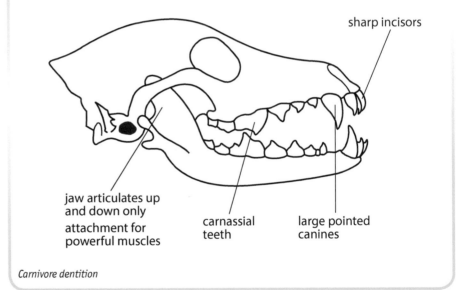

sharp incisors

jaw articulates up and down only
attachment for powerful muscles

carnassial teeth

large pointed canines

Carnivore dentition

Herbivores

The difficulty in eating a vegetative diet is that the main component of cell walls in plant tissue is cellulose. Whilst it is made up of β-glucose, the arrangement of these molecules into microfibrils makes cellulose very difficult to digest. Herbivores possess incisors and canine teeth that slice through vegetation cropping it. Some herbivores lack incisors in the upper jaw, instead possessing a horny pad, which the lower teeth cut against. A gap called the diastema allows food to be mixed during the chewing process. They possess interlocking molar teeth that are rough due to the presence of sharp enamel ridges. The teeth are worn down by the abrasive plant material and therefore grow continuously. The jaws are able to move in a sideways motion to aid the grinding of the food.

quickfire

③ Describe two differences between the teeth of a carnivore and those of a herbivore.

Sheep

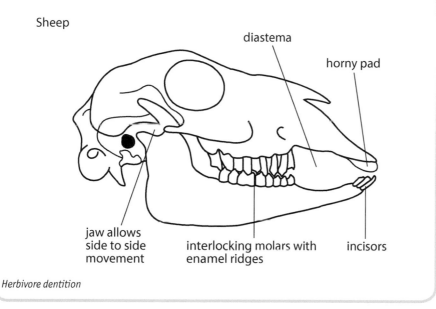

Herbivore dentition

Ruminants

Ruminants, including cows and sheep, possess a highly modified oesophagus containing three chambers, one of which is the rumen, and a 'true' stomach.

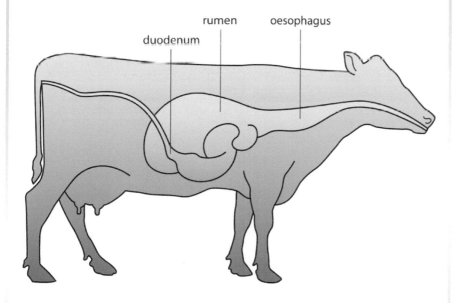

Ruminant gut

- Grass is first chewed to form a bolus known as the 'cud', which is swallowed and enters the rumen where it mixes with mutualistic cellulose-digesting bacteria that produce glucose from the cellulose. The glucose is anaerobically respired by bacteria producing organic acids, carbon dioxide and methane as waste products.

Grade boost

Don't say ruminants have four stomachs – they have four chambers, one of which is the 'true' stomach.

quickfire

④ Why do ruminants need cellulose-digesting bacteria?

≫ Pointer

Rabbits do not have a rumen. Instead, cellulose-digesting bacteria are found in the caecum and appendix. Therefore, they have to re-eat their faeces to absorb the products of cellulose digestion.

- Remaining grass passes to the reticulum where it is reformed into a cud, which is regurgitated, re-chewed to increase the surface area for the action of the bacterial cellulases before being swallowed again.
- The cud now passes to the omasum, where organic acids are absorbed into the blood.
- Finally the material passes to the abomasum (true stomach) where acid kills the bacteria and pepsin begins the digestion of the bacteria.
- Water is absorbed in the large intestine in a similar way as in humans.

Parasites

Parasites live in (endoparasites) or on (ectoparasites) a host organism, causing it harm. The pork tapeworm, *Taenia solium* has two hosts: the primary host (the host in which sexual reproduction takes place) is the human and the pig is the secondary host. Both are needed to fully complete the parasite's life cycle. If eggs are eaten directly by a human, rather than eating infected meat, cysts can form in the brain, causing a far more serious condition.

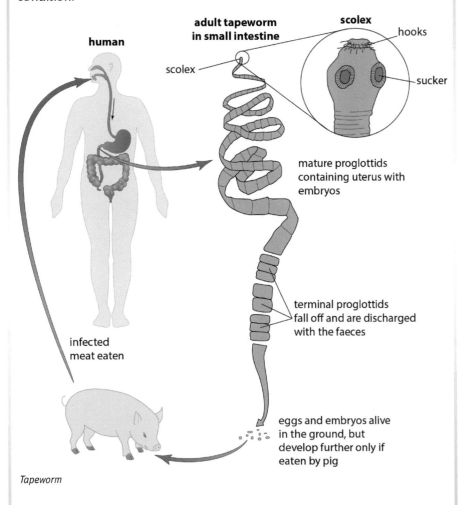

Tapeworm

The pork tapeworm has evolved many adaptations to overcome the harsh environment inside the human digestive system:

Adaptation	Reason
Suckers and hooks	Attachment to gut wall
Thin and large surface area to volume ratio	Maximise absorption of digested food
Produces enzyme inhibitors	Prevents digestion by host's enzymes
Thick cuticle	Protects it from host's immune responses
Has male and female reproductive structures	Allows for sexual reproduction without second tapeworm
Produces vast numbers of eggs (one segment can have as many as 125,000 eggs. Six segments pass out in the faeces per day)	Increases chances of finding another host
Eggs have resistant shells	Can survive until eaten by secondary host

The tapeworm does not have a digestive system, so absorbs the products of the host's digestion directly through its cuticle. The head louse (*Pediculus*) is an example of an ectoparasite, which feeds by sucking blood from the scalp of the host. It is a wingless insect and so can only pass to a new host via direct contact. It takes about 2 weeks for an egg to hatch into a nymph, which then feeds on blood. After a further 10 days, nymphs develop into adults, which can then lay eggs, and the cycle continues. It has evolved a number of adaptations to its parasitic mode of life:

- Legs are adapted to be claw-like to grip onto the hairs.
- Lays eggs that are glued to the base of hairs.

quickpire

(5) Name three problems that a tapeworm has to overcome.

Grade boost

Be prepared to explain how each adaptation helps the parasite.

Component 2 Summary

Classification and biodiversity

- Classification is hierarchical and organises similar organisms into domains, kingdoms, phyla, classes, orders, families, genera and species
- Relatedness of organisms can be investigated using physical features, DNA and amino acid sequences and immunology
- Biodiversity can vary over time and spatially
- Adaptation to different environments by natural selection

Adaptations for gas exchange

- Problems associated with increase in size – an organism's requirements increases by the X^3, availability via the X^2 therefore larger organisms need a specialised gas exchange surface e.g. gills or lungs, a ventilation mechanism, and a transport system. Unicellular organisms use simple diffusion
- Gas exchange in fish via gills, using parallel flow of water and blood in sharks and counter-current flow in bony fish. Gills are ventilated by movements of operculum
- Humans exchange gases via alveoli in lungs. Lungs are ventilated by negative pressure breathing
- Insects have a tracheal system of chitin lined tubes that end in tracheoles that come into direct contact with the tissues. Spiracles can close to reduce water loss
- Plants exchange gases via stomata which close at night or during times of water shortage

Adaptations for transport in animals

- Circulatory systems are either open or closed and use a pump. In animals, the main vessels are arteries, arterioles, capillaries, venules and veins
- Heart is myogenic and is controlled by specialised regions of cardiac fibres. The cardiac cycle involves atrial systole, ventricular systole and total diastole
- Blood consists of plasma, red blood cells containing haemoglobin (transport of oxygen), and white blood cells (body defence). Oxygen is transported as oxyhaemoglobin. Carbon dioxide is transported largely as hydrogen carbonate ions
- Formation and role of tissue fluid in supplying nutrients to cells and removing waste

Adaptations for transport in plants

- Distribution of vascular tissue in plants is different in roots, stems and leaves. Xylem transports water and dissolved mineral ions whilst phloem sieve tubes transport solutes (sucrose and amino acids) from SOURCE to SINK via mass flow theory
- Water moves through the root via apoplast, symplast and vacuolar pathways
- Transpiration is the loss of water vapour from the leaf via stomata. Forces involved include: adhesive and cohesive forces, root pressure and transpiration pull
- Plants have become adapted to living in dry (xerophytes) and wet (hydrophytes) environments

Adaptations for nutrition

- Modes of nutrition – plants are autotrophic and can manufacture food by photosynthesis. Animals are heterotrophic and consume complex organic material
- Food is processed by ingestion, digestion, absorption and egestion. Different regions of the gut are specialised for these roles. The end products are glucose, amino acids, fatty acids and glycerol
- Adaptations to different diets – herbivores and carnivores have evolved different dentition. Ruminants have a specialised stomach in which cellulose digesting bacteria live
- Parasitism – parasites live in or on a host causing it harm. They have evolved adaptations to survive in hostile conditions, e.g. inside the human gut

Exam practice and technique

Aims and objectives

The Eduqas AS and A Levels in Biology aim to encourage learners to:

- Develop essential knowledge and understanding of different areas of biology and how they relate to each other.
- Develop and demonstrate a deep appreciation of the skills, knowledge and understanding of scientific methods used within biology.
- Develop competence and confidence in a variety of practical, mathematical and problem solving skills.
- Develop their interest in and enthusiasm for biology, including developing an interest in further study and careers associated with the subject.
- Understand how society makes decisions about biological issues and how biology contributes to the success of the economy and society.

Types of exam question

There are **two** main types of questions in the exam:

1. Short-answer structured questions

The majority of questions fall into this category. These questions may require description, explanation, application, and/or evaluation, and are generally worth 6-10 marks. Application questions could require you to use your knowledge in an unfamiliar context or to explain experimental data. The questions are broken down into smaller parts e.g. a), b), c), etc., which can include some 1-mark name or state questions. You could also be asked to complete a table, label or draw a diagram, plot a graph, or perform a mathematical calculation.

Some examples requiring name or state:

- Name the bond labelled X on the diagram.   (1 mark)
- Name the type of cell division taking place.   (1 mark)
- What is the name given to this type of diagram?   (1 mark)

Some examples requiring mathematical calculation:

- The magnification of the image above is × 32 500. Calculate the actual width of the organelle in micrometres between points A and B.   (2 marks)
- Using the formula and the table given below, calculate the Diversity index.   (3 marks)

Some examples requiring description:

- Describe the results for the free enzymes at temperatures above 40°C. (2 marks)
- Describe one similarity and one difference between the structures of chitin and cellulose. (2 marks)
- Describe how a sweep net could be used to estimate the Diversity index of insects at the base of a hedge. (3 marks)

Some examples requiring explanation:

- Explain why there must be three bases in each codon to assemble the correct amino acid. (2 marks)
- Giving examples explain the difference between homologous and analogous structures. (2 marks)
- Explain how the structures of cellulose and chitin are different from that of starch. (2 marks)

Some examples requiring application:

- Using your knowledge of the structure of cell membranes, explain why ethanol causes the red pigment betacyanin to leak out of the beetroot cells. (2 marks)
- Explain why the rate of water uptake in the plant increased as the wind speed increased. (3 marks)
- What conclusions could be drawn from this experiment regarding the effect of temperature on cell membranes? (3 marks)

Some examples requiring evaluation:

- Describe how you could improve your confidence in your conclusion. (2 marks)
- Comment on the validity of your conclusion. (2 marks)
- Evaluate the strength of their evidence and hence the validity of their conclusion. (4 marks)

2 Longer essay questions

As part of the exam, you will need to answer an extended response question worth 9 marks. The quality of your extended response will be assessed in this question. You will be awarded marks based upon a series of descriptors: to gain the top marks it is important to give a full and detailed account including a detailed explanation. You should use scientific terminology and vocabulary accurately, including accurate spelling and use of grammar and include only relevant information. It is a good idea to do a brief plan before you start to organise your thoughts: You should cross this out once you have finished. We will look at some examples later.

Command or action words

These tell you what you need to do. Examples include:

Analyse means to examine the structure of data, graphs or information. A good tip is to look for trends and patterns, and maximum and minimum values.

Calculate is to determine the amount of something mathematically. It is really important to show your working (if you don't get the correct answer you can still pick up marks for your working).

Choose is to select from a range of alternatives.

Compare involves you identifying similarities and differences between two things. It is important when detailing similarities and differences that you talk about both. A good idea is to make two statements, linked with the word *'whereas'*.

Consider is to review information and make a decision.

Complete means to add the required information.

Describe means give an account of what something is like. If you have to describe the trend in some data or in a graph then give values.

Discuss involves presenting the key points.

Distinguish involves you identifying differences between two things.

Draw is to produce a diagram of something.

Estimate is to roughly calculate or judge the value of something.

Evaluate involves making a judgment from available data, conclusion or method, and proposing a balanced argument with evidence to support it.

Explain means give an account and use your biological knowledge to give reasons why.

Identify is to recognise something and be able to say what it is.

Justify is about you providing an argument in favour of something; for example, you could be asked if the data support a conclusion. You should then give reasons why the data supports the conclusion given.

Label is to provide names or information on a table, diagram or graph.

Outline is to set out the main characteristics.

Name means identify using a recognised technical term. Often a one-word answer.

State means give a brief explanation.

Suggest involves you providing a sensible idea. It is not straight recall, but more about applying your knowledge.

General exam tips

Always read the question carefully: read the question twice! It is easy to provide the wrong answer if you don't give what the question is asking for. All the information provided in the question is there to help you to answer it. The wording has been discussed at length by examiners to ensure that it is as clear as possible.

Look at the number of marks available. A good rule is to make *at least* one different point for each mark available. So make five different points if you can for a four-mark question to be safe. Make sure that you keep checking that you are actually answering the question that has been asked – it is easy to drift off topic!

If a diagram helps, include it: but make sure it is fully annotated.

Timing

On the AS Eduqas exam there are 75 marks for each AS exam paper and you have 90 minutes. This gives you a sense of how much time you should spend on each exam question, and a good rule is about one mark per minute. Don't forget that this timing is not just about writing but you should spend time thinking, and for the extended answer some planning, too. At A level there are three component exams, each worth 100 marks and you have 2 hours. The A level exam tests your knowledge on BOTH years of the course.

Assessment objectives

Examination questions are written to reflect the assessment objectives (AOs) as laid out in the specification. The three main skills that you must develop are:

AO1: Demonstrate knowledge and understanding of scientific ideas, processes, techniques and procedures.

AO2: Apply knowledge and understanding of scientific ideas, processes, techniques and procedures.

AO3: Analyse, interpret and evaluate scientific information, ideas and evidence, including in relation to issues.

In both written examinations you will also be assessed on your:

- mathematical skills (minimum of 10%)
- practical skills (minimum of 15%)
- ability to select, organise and communicate information and ideas coherently using appropriate scientific conventions and vocabulary.

In any one question, you are likely to be assessed on all skills to some degree. It is important to remember that only about a third of the marks are for direct recall of facts. You will need to apply your knowledge, too. If this is something you find hard, practise as many past paper questions as you can. Many examples come up in slightly different forms from one year to another.

Your practical skills will be developed during class time sessions and will be assessed in the examination papers. This could include:

- plotting graphs
- identifying controlled variables and suggesting appropriate control experiments
- analysing data and drawing conclusions
- evaluating methods and procedures and suggesting improvements.

Drawing graphs

Full marks are rarely awarded for graphs. Common errors include:

- incorrect labels on axes
- missing units
- sloppy plotting of points
- failing to join plots accurately
- non-linear scales.

Can you spot the mistakes?

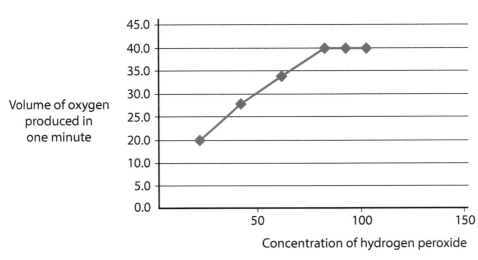

Mistakes are:

- No units on either axis.
- No value for origin on horizontal axis.
- Vertical axis is nonlinear, i.e. gaps are unequal.

Make sure that you draw range bars and can explain their significance.

Understanding AO1: Demonstrate knowledge and understanding

You will need to demonstrate knowledge and understanding of scientific ideas, processes, techniques and procedures.

36% of the questions set on the Eduqas AS exam papers and 30% of the questions set on the A Level Eduqas exam papers are for demonstrating knowledge and understanding.

Common command words used here are: state, name, describe, explain.

This involves recall of ideas, processes, techniques and procedures detailed in the specification. This is content you should know.

A good answer is one that uses detailed biological terminology accurately, and has both clarity and coherence.

If you were asked to describe the structure of starch and explain how its structure makes it a good molecule for storage of glucose, you might write:

'Starch is made from alpha glucose molecules joined together. Because it is insoluble it is ideal for storage.'

This is a basic answer.

A good answer needs to be both accurate and detailed. For example, *'Starch is a polymer of alpha glucose molecules joined by condensation reactions. It consists of two molecules: amylose which is a straight chain formed from 1-4 glycosidic bonds and amylopectin which is highly branched formed from alpha glucose joined by 1-4 glycosidic bonds with 1-6 branches every 20 or so glucose molecules. Starch is very compact, and because amylose and amylopectin are both insoluble it does not affect the water potential of the cell, and so makes an ideal storage molecule in plant cells. Amylopectin releases glucose very quickly as there are many ends on which the enzyme amylase can act.'*

Understanding AO2: Applying knowledge and understanding

You will need to apply knowledge and understanding of scientific ideas, processes, techniques and procedures:

- in a theoretical context
- in a practical context
- when handling qualitative data (this is data with no numerical value, e.g. a colour change)
- when handling quantitative data (this is data with a numerical value, e.g. mass/g).

44% of the questions set on the Eduqas AS exam papers and 45% of the questions set on the A Level Eduqas exam papers are for application of knowledge and understanding.

Common command words used here are: describe (if its unfamiliar data or diagrams), explain and suggest.

This involves applying ideas, processes, techniques and procedures detailed in the specification to unfamiliar situations.

Describing data

It is important to describe accurately what you see, and to quote data in your answer.

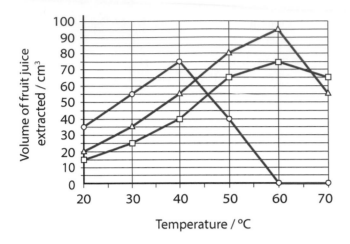

- Free enzymes
- Enzyme bound to gel membrane surface
- Enzyme immobilised inside beads

If you were asked to compare the volume of juice produced when using enzymes bound to the gel membrane surface compared to the enzyme immobilised inside the beads, you might write:

'The volume of juice extracted increases with temperature up to the optimum temperature of 60°C in both enzymes. Above this, the volume of juice decreases.'

This is a basic answer.

A good answer needs to be both accurate and detailed. For example,

'Increasing temperature causes the volume of fruit juice extracted to increase up to 60°C. The volume of juice collected is higher up to 60°C with the enzyme bound to the gel membrane, peaking at 95cm³ compared to 75cm³ for the enzyme immobilised inside the beads. Above 60°C the volume of fruit juice extracted decreases, but this is more noticeable for the enzymes bound to the gel membrane surface which decrease by 40cm³ compared to just 10cm³ for the enzyme immobilised inside the beads.'

If you were also asked to explain the results, a basic answer would include reference to *'increased kinetic energy up to 60°C, and denaturing enzymes above 60°C'*. A good answer is one that uses detailed biological terminology accurately, and has clarity and coherence. A good answer would also include reference to *'increased enzyme–substrate complexes forming up to 60°C'* and would include that *'above 60°C, hydrogen bonds break resulting in the active site changing shape so fewer enzyme–substrate complexes could form'*.

Mathematical requirements

A minimum of 10% of marks across the whole qualification will involve mathematical content. Some of the mathematical content requires the use of a calculator, which is allowed in the exam. In the specification it states that calculations of the mean, median, mode and range may be required, as well as percentages, fractions and ratios. Some additional requirements are included at A level **which are shown in bold**.

You will be required to process and analyse data using appropriate mathematical skills. This could involve considering margins of error, accuracy and precision of data.

Concepts	Tick here when you are confident you understand this concept
Arithmetic and numerical computation	
Convert between units, e.g. mm^3 to cm^3	
Use an appropriate number of decimal places in calculations, e.g. for a mean	
Use ratios, fractions and percentages, e.g. calculate percentage yields, surface area to volume ratio	
Estimate results	
Use calculators to find and use power, exponential and logarithmic functions, e.g. estimate the number of bacteria grown over a certain length of time	
Handling data	
Use an appropriate number of significant figures	
Find arithmetic means	
Construct and interpret frequency tables and diagrams, bar charts and histograms	
Understand the principles of sampling as applied to scientific data, e.g. use Simpson's Diversity Index to calculate the biodiversity of a habitat	
Understand the terms mean, median and mode, e.g. calculate or compare the mean, median and mode of a set of data, e.g. height/mass/size of a group of organisms	
Use a scatter diagram to identify a correlation between two variables, e.g. the effect of lifestyle factors on health	
Make order of magnitude calculations, e.g. use and manipulate the magnification formula : magnification = size of image / size of real object	
Understand measures of dispersion, including standard deviation and range	
Identify uncertainties in measurements and use simple techniques to determine uncertainty when data are combined, e.g. calculate percentage error where there are uncertainties in measurement	
Algebra	
Understand and use the symbols: $=, <, <<, >>, >, \propto, \sim$.	
Rearrange an equation	
Substitute numerical values into algebraic equations	
Solve algebraic equations, e.g. solve equations in a biological context, e.g. cardiac output = stroke volume × heart rate	
Use a logarithmic scale in the context of microbiology, e.g. growth rate of a microorganism such as yeast	
Graphs	
Plot two variables from experimental or other data, e.g. select an appropriate format for presenting data	
Understand that $y = mx + c$ represents a linear relationship	
Determine the intercept of a graph, e.g. read off an intercept point from a graph, e.g. compensation point in plants	
Calculate rate of change from a graph showing a linear relationship, e.g. calculate a rate from a graph, e.g. rate of transpiration	
Draw and use the slope of a tangent to a curve as a measure of rate of change	
Geometry and trigonometry	
Calculate the circumferences, surface areas and volumes of regular shapes, e.g. calculate the surface area or volume of a cell	

Understanding AO3: Analysing, interpreting and evaluating scientific information

This is the last and most difficult skill. You will need to analyse, interpret and evaluate scientific information, ideas and evidence, to:

- make judgements and reach conclusions
- develop and refine practical design and procedures.

20% of the questions set on the Eduqas AS exam papers and 25% of the questions set on the A Level Eduqas exam papers are for analysing, interpreting and evaluating scientific information.

Common command words used here are: evaluate, suggest, justify and analyse.

This could involve:

- Commenting on experimental design and evaluating scientific methods.
- Evaluating results and draw conclusions with reference to measurement, uncertainties and errors.

What is accuracy?

Accuracy relates to the apparatus used: How precise is it? What is the percentage error? For example, a 5ml measuring cylinder is accurate to ± 0.1 ml so measuring 5ml could yield 4.9 – 5.1 ml. Measuring the same volume in a 25ml measuring cylinder which is accurate to ± 1 ml would yield 4-6ml.

Calculating % error

It's a simple equation: accuracy/starting amount × 100. For example, in the 25ml measuring cylinder the accuracy is ± 1 ml so the error is $1/25 \times 100 = 4\%$, whereas in the 5ml cylinder the accuracy is ± 0.1 ml so the error is $0.1/5 \times 100 = 2\%$. Therefore, for measuring 5mls it is best to use the smaller cylinder as the % error is lowest.

What is reliability?

Reliability relates to your repeats. In other words, if you repeat the experiment three times and the values obtained are very similar, then it indicates that your individual readings are reliable. You can increase reliability by ensuring that all variables that could influence the experiment are controlled, and that the method is consistent.

Describing improvements

If you were asked to describe what improvements could be made to the reliability of the results obtained from an experiment extracting apple juice, you would need to look closely at the method and apparatus used.

Q: Pectin is a structural polysaccharide found in the cell walls of plant cells and in the middle lamella between cells, where it helps to bind cells together. Pectinases are enzymes that are routinely used in industry to increase the volume and clarity of fruit juice extracted from apples. The enzyme is immobilised onto the surface of a gel membrane, which is then placed inside a column. Apple pulp is added at the top, and juice is collected at the bottom. The process is shown in the diagram. Describe what improvements could be made.

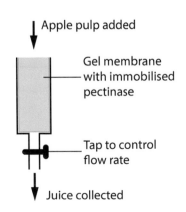

Apple pulp added

Gel membrane with immobilised pectinase

Tap to control flow rate

Juice collected

You might write:

'I would make sure that the same mass of apples is added, and that they were the same age.'

This is a basic answer.

A good answer needs to be both accurate and detailed. For example,

'I would make sure that the same mass of apples is added, for example 100g, and that they were the same age, e.g. 1 week old. I would also control the temperature at an optimum for the pectinases involved, e.g. 30°C.'

Look at the following example:

A student carried out an experiment to investigate the effect of temperature on cell membranes. Using a borer, equal sized pieces of beetroot were cut, washed, and blotted with a paper towel. Each piece was placed into a test tube containing 25cm³ of 70% ethanol (an organic solvent) and incubated at 15°C. A red pigment called betacyanin found in the vacuoles of the beetroot cells began to leak out into the ethanol turning it red. The experiment was repeated at 30°C and 45°C and the time taken for the ethanol to turn red was recorded below.

Temperature (°C)	Time taken for the ethanol to turn red (s)			
	Trial 1	Trial 2	Trial 3	Mean
15	450	427	466	447.7
30	322	299	367	329.3
45	170	99	215	161.3

Q: What conclusions could be drawn from this experiment regarding the effect of temperature on cell membranes?

You might write:

'Increasing temperature increases the amount of dye that leaks from the cells.'

A good answer needs to be both accurate and detailed. For example,

'Increasing temperature increases the kinetic energy of the membrane and dye molecules. The increased movement of membrane molecules increases the number of gaps in membrane so more dye can escape from the cells.'

If asked to comment on the validity of your conclusion, you might write:

'It was difficult to determine when the solutions turned red, making it difficult to know when to stop timing the reactions.'

A good answer would be more detailed, for example,

'The results at 45°C are very variable and range from 99 to 215 seconds. It is difficult to reach a conclusion about the effect of temperature on cell membranes as only three temperatures were investigated. Another major difficulty would be in determining the end point of the reaction, as no standard red colour was used.'

Questions and answers

This part of the guide looks at actual student answers to questions. There is a selection of questions covering a wide variety of topics. In each case there are two answers given; one from a student (Isla) who achieved a high grade and one from a student who achieved a lower grade (Ceri). We suggest that you compare the answers of the two candidates carefully; make sure you understand why one answer is better than the other. In this way you will improve your approach to answering questions. Examination scripts are graded on the performance of the candidate across the whole paper and not on individual questions; examiners see many examples of good answers in otherwise low scoring scripts. The moral of this is that good examination technique can boost the grades of candidates at all levels.

Component 1

Component 2

Q & A 1

(a) Name the monomer that makes up the polymer, and its form. [1]

(b) Name the bond formed between the two hexose sugars. [1]

(c) State one structural difference between this molecule and cellulose. [1]

(d) Explain how the molecule shown gives strength to the exoskeleton. [2]

Isla's answer

a) β glucose ✓

b) 1–4 glycosidic bond ✓

c) some OH groups have been replaced with $NHCOCH_3$ ✓

d) hydrogen bonds form between the straight chains of β glucose molecules which then form microfibrils. ✓ ①

Examiner commentary

① Isla's answer should have been expanded to include that adjacent molecules are rotated by 180° and that hydrogen bonds form between parallel chains.

Isla achieves 4/5 marks

Ceri's answer

a) glucose ✗ ①

b) 1–4 glucosidic bond ✗ ②

c) there are $NHCOCH_3$ groups present ✗ ③

d) chains form microfibrils ✗ ④

Examiner commentary

① Ceri has not named the form, β.

② Ceri has misspelled glycosidic which could be confused with other terms.

③ Ceri has identified the additional group but has not compared it with cellulose by saying that the OH groups have been replaced.

④ Ceri did not explain how the chains form microfibrils.

Ceri achieves 0/5 marks

Exam tip

It is important that you read the question carefully and give as detailed a response as you can. Spelling is important especially if you write a word that could be confused with something else, in this case glucose. When you are asked to give a difference, you must say something about **both** – in this case cellulose **and** chitin.

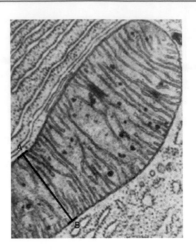

(a) The nucleus has pores in the envelope that surrounds it, whereas the organelle shown does not. Name the organelle shown and describe **one** *other* difference between the membranes that surround the organelle shown and the nucleus. [2]

(b) The surface area of the organelle shown can be calculated by using the formula **surface area = $2\pi r l + 2\pi r^2$**, where l = length of the organelle is 9.1μm, π = 3.14, diameter is 1.0μm. **Estimate the surface area of the organelle shown. Show your working.** [3]

(c) The surface area of a spherical organelle with the **same** volume is 23.1μm². Results from experiments have concluded that there are advantages to the cell of these organelles possessing a cylindrical shape. Evaluate this statement. [4]

Isla's answer

a) Mitochondrion. ✓ Mitochondria have a folded inner membrane which the nucleus doesn't ✓ ①

b) Radius = 0.5 ✓
$(2 \times 3.14 \times 0.5 \times 9.1) + (2 \times 3.14 \times 0.25)$ ✓
= 30.14 (μm²) ✓

c) A spherical mitochondrion will have a larger diffusion distance to the centre than a cylindrical one ✓ and its surface area: volume ratio will be higher. ✓ This results in more oxygen being absorbed and carbon dioxide lost by diffusion. ✓ ②

Examiner commentary

① Isla makes a good comparison between the membranes in both organelles. Another point that Isla could have made is that nuclear membranes have ribosomes attached, whilst mitochondrial membranes don't.

② Isla gave a good answer, but a perfect answer would have made reference to more efficient respiration and more ATP produced as a result.

Isla achieves 8/9 marks

Ceri's answer

a) Mitochondria. ✓ They have a folded inner membrane ①

b) Radius = 0.5 ✓
= 31.71 (μm²) ✗ ②

c) A spherical mitochondrion will have a larger diffusion distance to the centre than a cylindrical one ✓ which results in more oxygen being absorbed. ③

Examiner commentary

① Ceri incorrectly used the plural for mitochondrion, but was still awarded the mark. Ceri did not however make a comparison between both organelles.

② Because Ceri shows limited working, only one mark can be awarded for correct calculation of radius.

③ Ceri correctly identifies the effect on the diffusion distance and hence absorption of oxygen, but the method of absorption (diffusion) should have been included.

Ceri achieves 3/9 marks

Exam tip

ALWAYS show your working with maths questions. It allows examiners to award some marks for process, and allows error carried forward (this is where if all other working is correct but you use an incorrect value from earlier, you can still pick up some marks). After doing a calculation ask yourself if it is a reasonable answer and that you have got the decimal point in the correct place, converted one unit to another correctly, put the correct units.

The diagram below shows a component of DNA.

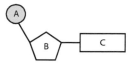

a) Name the molecules A, B and C. [2]
b) Describe two differences between a polymer of DNA and a polymer of RNA. [2]
c) The table below shows the bases guanine and cytosine as percentages of the total nucleotides present in three different micro-organisms that were calculated by sequencing the genome of each organism. The scientists concluded that the virus contained single stranded RNA. What evidence is there to support this conclusion? [2]

Micro-organism	Base composition (%)	
	G	C
yeast	18.7	17.1
bacteria	36.0	35.7
virus	42.0	13.9

Isla's answer

a) A= phosphate ✓
   B= deoxyribose ✓
   C= nitrogenous base ✓   all correct = 2 marks
b) uracil replaces thymine ✓, and RNA is usually single stranded, whereas DNA is double stranded. ✓
c) In double stranded molecules the proportion of complimentary bases G and C must be equal. In the virus 42% is guanine and 13.9% is cytosine so molecule can't be double stranded, it must be single stranded. ✓   ①

Examiner commentary

① Isla correctly states that guanine and cytosine aren't equal so the molecule can't be stranded. To gain full marks, Isla should explain that there is no evidence to support that the molecule is RNA, as there is no reference to the proportion of thymine (which would be nil in RNA).

Isla achieves 5/6 marks

Ceri's answer

a) A = phosphate ✓
   B = pentose ✗
   C = nitrogenous base ✓   ① two correct
                                = 1 mark
b) there is no thymine, and RNA is usually single stranded ✗   ②
c) There is no thymine present so it must be RNA which is usually single stranded. ✗   ③

Examiner commentary

① Ceri fails to name the pentose sugar in DNA.
② Ceri identifies that there is no thymine present and that RNA is single stranded, but no comparison between DNA and RNA has been made.
③ No evidence is provided, just a statement based upon Ceri's knowledge of RNA, which was assessed in part b).

Ceri achieves 1/6 marks

> **Exam tip**
>
> It is important when asked to *Justify* or provide evidence for a conclusion that you evaluate any data provided. You should use this to support your answer.

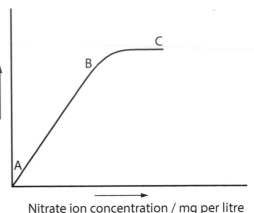

(a) Water enters root hair cells by osmosis. The water potential (Ψ) of the soil water is Ψ −100kPa, and the pressure potential (Ψ_P) inside the root hair cell is +200kPa. Calculate the solute potential (Ψ_S) of the root hair cell if there is no net movement of water using the formula $\Psi = \Psi_S + \Psi_P$. Show your working. [2]

b) An experiment was carried out to determine how nitrate ions enter the roots of plants. The results are shown opposite.

(i) What conclusion can you draw from the evidence provided as to how nitrate ions are absorbed? [2]

(ii) What further evidence would be needed to confirm exactly which method of uptake is involved? [2]

Isla's answer

a) $\Psi_S = \Psi - \Psi_P$ / i.e. −100 −200 ✓
   = −300 ✗ ①

b) (i) Nitrate ions use a carrier protein ✓ because above point B, further increase in ion concentration has no effect on the rate of uptake ✓ ②

(iii) I would repeat the experiment in the absence of oxygen to see if uptake stopped. ✓ This would confirm that it is active transport as oxygen is needed to produce ATP rather than facilitated diffusion ✓ ③

Ceri's answer

a) $\Psi_S = \Psi - \Psi_P$ / i.e. −100 −200 ✓
   = −300 kPa; ✓

b) (i) Nitrate ions are absorbed by diffusion? ✗ because it is directly affected by concentration of nitrate ions between A and B? ✗ ①

(ii) I would repeat the experiment ✗ ②

Examiner commentary

① Isla did not include units for her part a) answer so lost one mark.

② Isla could have included in i) that this is because all the channel proteins are in use.

③ Part ii) is a perfect answer: It names both methods and a sensible idea of how to distinguish between them based upon the knowledge that active transport requires ATP.

Isla achieves 5/6 marks

Examiner commentary

① Whilst it is true that rate of diffusion is proportional to concentration, this explanation fails to explain what happens at B – C.

② Repeating the experiment in part ii) would just provide more evidence and so improve reliability of the conclusion.

Ceri achieves 2/6 marks

Exam tip

Repeating an experiment only improves reliability: it does not give more evidence.

The diagram on the right shows the relative lengths of the cell cycle in actively dividing cells taken from the root tip of a garlic plant.

(a) Describe the changes that occur to the nucleus of a plant cell during prophase. [3]

(b) A new drug has been developed that inhibits mitosis by preventing the formation of the spindle fibres. Garlic bulbs were grown in a solution of the new drug and the quantity of DNA present in a cell from the root tip was measured over the 24-hour length of the cell cycle. The results are shown below together with the results from garlic bulbs grown in water.

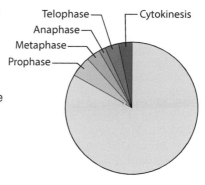

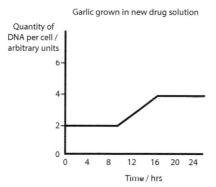

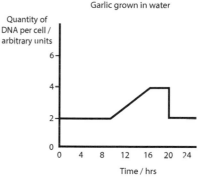

Using your knowledge of the cell cycle, explain how the results of this experiment show that the new drug inhibits mitosis. [3]

Isla's answer

a) During prophase the nuclear membrane disappears ✓ and the chromosomes condense appearing as two distinct chromatids ✓ joined at centromere. ✓

b) Interphase occurs in both experiments as DNA quantity doubles. ✓ When placed in the new drug, there is no halving of DNA at 20 hours as seen in the garlic grown in water. ✓ This shows that there is no cytokinesis because the lack of spindle fibres prevents anaphase. ✓

Examiner commentary

Isla has learnt the key events in prophase and was able to relate the graph to her knowledge of cytokinesis.

Isla achieves 6/6 marks

Ceri's answer

a) During prophase the chromosomes condense ✓ and centrioles appear. ✗ ①

b) When placed in the drug, DNA does not decrease, ✗ so cell division has not happened. ✗ ②

Examiner commentary

① Reference to centrioles appearing is incorrect as they are absent in higher plants.

② Reference to DNA decreasing is too vague. The question asks how the results show that the new drug inhibits mitosis: references to cell division not occurring is also too vague, when information in the question details that the new drug prevents spindle formation. Ceri should link this to anaphase not occurring, and that subsequently cytokinesis will not occur.

Ceri achieves 1/6 marks

Exam tip

Learn the key events in mitosis (and meiosis) and be able to recognise drawings of each stage. You should be able to use data given to reach or support a conclusion.

Pectin is a structural polysaccharide found in the cell walls of plant cells and in the middle lamella between cells, where it helps to bind cells together. Pectinases are enzymes that are routinely used in industry to increase the volume and clarity of fruit juice extracted from apples. The enzyme is immobilised onto the surface of a gel membrane, which is then placed inside a column. Apple pulp is added at the top, and juice is collected at the bottom. The process is shown in the diagram.

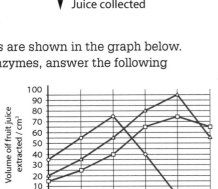

a) Explain why reducing the flow rate of material through the column would result in increased juice collected. [1]

b) The extraction of juice using pectinase was compared using equal volumes and concentrations of free enzyme, enzymes bound to the surface of a gel membrane and enzymes encapsulated inside alginate beads. The results are shown in the graph below.

Using the graph and your own knowledge of enzymes, answer the following questions.

   i) Describe and explain the results for the free enzymes at temperatures above 40°C [4]

   ii) Explain why a higher yield of juice was obtained when using free enzymes between temperatures of 20°C and 40°C than when using either kind of immobilised enzyme. [2]

   iii) Explain the differences seen in the results for the enzymes bound to the gel membrane surface with those immobilised inside the beads, between temperatures of 20°C and 60°C. [2]

c) Name two variables other than flow rate that must be controlled in this experiment and explain the effect of not controlling them. [3]

Isla's answer

a) Allows more time for the pectinase to break down the apple into juice and so more enzyme substrate complexes are formed. ✓

b) i) Above 40°C less juice is extracted, ✓ above 60°C no juice is extracted ✓ because at 60°C the enzymes are fully denatured ✓ due to the hydrogen bonds breaking. ①

ii) The free enzymes can move about and so have more kinetic energy ✓ and so are more likely to collide with the pectinase to form ES complexes. ✗ ②

iii) More juice is collected when membrane bound enzymes are used because they directly touch the fruit. ✓ ③ ✓

c) You should use apples of the same age ④. The pH of apples should also be kept constant. ✓ ⑤. If the pulp was too acidic then the pectinase enzymes would become inactivated or denatured which would reduce the yield of fruit juice. ✓

Examiner commentary

Isla shows a good understanding of a complex application of enzyme kinetics and immobilised enzymes to a practical situation.

① Isla could have explained the effect upon the active site of the hydrogen bonds breaking i.e. that the tertiary structure deforms.

② Isla uses the abbreviation ES – this is not an accepted abbreviation for enzyme–substrate.

③ Isla needed to explain why less juice would be extracted from enzymes immobilised inside the beads i.e. that the substrate has to diffuse into the bead.

④ Isla should include a reason why this would affect the yield, for example as riper apples may yield more juice as cell walls begin to break down naturally with age.

⑤ Isla gains one mark for two correct controlled variables, and explains why it is important to control pH. Another variable to control here would be concentration of enzyme.

Isla achieves 9/12 marks

Ceri's answer

a) There is more time for the pectinase to digest the apple. ✗ ①

b) i) Above 40°C less juice is extracted, ✓ because the enzymes are denatured ✗, because the peptide bonds break. ✗ ②

ii) The free enzymes can move about and therefore have more kinetic energy ✓ so there are more collisions between the enzyme and substrate. ✗ ③

iii) More juice is collected from the membrane bound enzymes. ✗ ④

c) Temperature should be kept constant, because at higher temperatures the high kinetic energy denatures the enzyme's active site. ✗ ⑤. pH should also be kept at the optimum for the enzyme say at pH 7. ⑥

Examiner commentary

① Need to explain fully why more juice is extracted, including reference to enzyme–substrate complexes.

② It is not correct to say at 50°C that the enzymes are denatured as some juice is still being extracted. Many are, but not all, so better to say above 40°C enzymes are denaturing. Ceri incorrectly says that peptide bonds break.

③ Ceri makes reference to more collisions but these must be **successful**, i.e. enzyme–substrate complexes form.

④ Ceri only describes. There is no explanation, so no marks can be awarded.

⑤ Temperature cannot be controlled in this experiment because it is an independent variable. The consequence on the active site of high temperature is explained, but not on the yield of juice, which was the dependent variable in the experiment.

⑥ Two correct controlled variables are needed for one mark, so as only one is correct, no mark is awarded. Ceri needs to explain the effect of not controlling pH on the yield of juice.

Ceri achieves 2/12 marks

Exam tip

Be careful with abbreviations – don't use ES complexes for enzyme–substrate complexes unless you have written it out in full first. It is vital that you read the question and follow the command words to ensure that you give the answer required and so pick up full marks. When you are required to explain in terms of enzyme kinetics, make sure you include reference to enzyme–substrate complexes.

Using examples, explain how the structure of carbohydrates and lipids enable them to perform their variety of functions in living organisms.   [9]

Isla's answer

Carbohydrates contain carbon, hydrogen and oxygen and are used in respiration, as storage molecules and provide structural support. Glucose is a six-carbon monosaccharide, which is the main source of energy for living organisms. It is easily hydrolysed during respiration producing ATP, ✓ and acts as a building block for more complex polysaccharides. Starch is used to store glucose in plants because unlike glucose it is insoluble and therefore osmotically inert. ✓ It is made from amylose and amylopectin. ① Amylose contains alpha glucose molecules joined via 1-4 glycosidic bonds into straight chains, which twist to form helices. In amylopectin the alpha glucose molecules are more branched, due to 1-4 and 1-6 glycosidic bonds. This creates a structure that is highly compact and so is easily stored within plant cells ✓ but it is easily hydrolysed into glucose when needed for respiration. ② In animals glucose is stored in the form of glycogen. Glycogen is another highly compact molecule that is osmotically inert, but it is made from a branched molecule containing alpha glucose joined by 1-6 glycosidic bonds, similar in structure to amylopectin. ✓

Plants and animals also use carbohydrates for structural support. Plant cells walls are strengthened by cellulose, which is made from chains of beta glucose molecules. Alternating glucose molecules rotate 180° and form straight chains. ✓ Hydrogen bonds then form between the long parallel chains forming microfibrils, which in turn are held together into fibres. ✓ The presence of fibres arranged at right angles to other fibres in the cell wall provides strength. ✓ In insects, chitin is formed in a similar way, except some OH groups are replaced by acetylamine groups. Their arrangement is similar, resulting in a strong and lightweight molecule found in the exoskeleton of insects. ✓   ③

Lipids also contain carbon, hydrogen, and oxygen, but the proportion of oxygen is less. They are non-polar molecules and hence insoluble in water, enabling their use as a waterproofing agent in the form of leaf waxes, and oils on bird's feathers. ✓ Their insolubility also makes lipids a good energy store in the form of oils in seeds and saturated fatty acids in animals. ✓ Storing fats under the surface of the skin also helps prevents heat loss as fats are poor conductors of heat, ✓ and as a component of myelin, they surround neurones providing electrical insulation. ✓ Due to the high numbers of hydrogen atoms, their hydrolysis releases more energy per gram than carbohydrates – in fact twice the energy, ✓ and also releases metabolic water, which is important in desert animals like the kangaroo rat where water is scarce. Some fats are stored around delicate organs like the kidney providing some protection against physical damage. Lipids are also key components in the plasma membrane of cells where they form phospholipids. ✓   ④

Carbohydrates and lipids are also needed to make other molecules important to plants and animals such as nucleotides with the addition of phosphate ✓ and chlorophyll with magnesium. ✓

Examiner commentary

① Isla describes the main functions of glucose well.

② The structure of starch has been described well to explain its role as a storage molecule.

③ The formation of microfibrils is clearly described and used to explain the role of chitin in structural support.

④ The properties of lipids have been used to explain their roles.

Summative comments

Isla would receive full marks (**9/9**) for an answer that fully addresses the question using good examples to link structure to function of lipids and carbohydrates with no irrelevant inclusions or significant omissions. Isla uses scientific conventions and vocabulary appropriately and accurately.

Ceri's answer

Carbohydrates are used in respiration, and as storage molecules and provide structural support. ✓ Starch is stored in plants because unlike glucose it is insoluble and compact and so is easily stored within plant cells. ✓ ① In animals glucose is stored in the form of glycogen. Plant cells walls contain cellulose, which is made from beta glucose molecules which are made into fibres in the cell wall provides strength. ②
Lipids are insoluble in water, so <u>make good energy</u> in animals ③. They have more energy than carbohydrates ④ and also release water, which is important in deserts. Fats are stored around delicate organs like the kidney providing protection ⑤. They also are found in the plasma membrane of cells where they form phospholipids.

Examiner commentary

① Ceri could have included that starch has no effect on osmosis in cells. Ceri would need to mention that starch is an energy/glucose store, and that the helical and branched structures of amylose and amylopectin make it more compact.

② It is not clear exactly what provides strength here. Ceri could have included detail on how beta glucose molecules are arranged into chains and microfibrils, and details on chitin

③ Whilst lipid insolubility is mentioned Ceri could link it to providing waterproofing.

④ Need to say twice as much energy.

⑤ Ceri needs to clarify what protection is conferred i.e. protection from physical damage.

Summative comments

Ceri would receive **1** out of **9** marks because whilst some relevant points are made showing limited reasoning, the structure of the key elements enabling them to perform their function is not explained. Ceri has made limited use of scientific conventions and vocabulary.

> **Exam tip**
> It is not one mark per point in the extended response, but more about how you answer the question. A plan is essential if you are to write coherently, and you must use scientific terminology correctly.

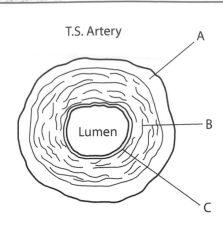

T.S. Artery

A

B

Lumen

C

(a) Label A, B and C shown in the diagram. [2]

(b) Patients suffering from angina complain of chest pains when performing relatively light exercise, because the oxygen supply to the heart muscles is reduced by a narrowing of the artery lumen caused by a build-up of fatty substances. The most common form of angina is stable angina, which is treated by using drugs which relax smooth muscles. Explain how the symptoms of angina are relieved. [2]

Isla's answer

a) A = collagen fibres ✓
   B = elastic muscle layer ✓
   C = endothelium ✓ three correct = 2 marks ①

b) Drugs relax the smooth muscles in layer B causing the artery lumen to widen ✓, this allows more blood to pass to the heart muscles supplying more oxygen ✓.

Examiner commentary

① Tunica externa/adventitia would also be acceptable for a), tunica media for b), tunica intima/ interna for c).

Isla achieves 4/4 marks

Ceri's answer

a) A = collagen ✓
   B = muscles ✓
   C = epithelium ✗ two correct = 1 mark ①

b) The drugs relax muscles meaning that more blood is supplied to the heart ✓ ②

Examiner commentary

① Whilst muscles is ok, muscle layer is better biology. Ceri has confused two similar sounding terms, epithelium and endothelium. Epithelium is a type of connective tissue not the layer shown.

② Ceri needs to relate the relaxed muscles to the structure of an artery, i.e. that the lumen widens.

Ceri achieves 2/4 marks

Q&A 9

The diagram below shows skulls from three different primates. *Australopithecus afarensis* and *Homo erectus* have been extinct for over a million years.

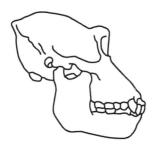
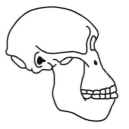
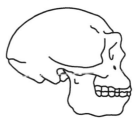

Gorilla gorilla   *Australopithecus afarensis*   *Homo erectus*

a) Name the class to which all these primates belong. [1]

b) Define the term species. [2]

c) (i) With reference to the diagrams suggest why scientists regard *Homo erectus* as being more closely related to *Australopithecus afarensis* than *Gorilla gorilla*. [1]

(ii) Using their classification, identify which primate is most closely related to modern humans, and explain your answer. [2]

Isla's answer

a) Vertebrata ✗ ①

b) A group of organisms with similar characteristics that can interbreed ✓ to produce fertile offspring. ✓

c) (I) The shape of the jaw and cranium of Homo erectus and Australopithecus afarensis look similar in shape. ✓

(ii) Homo erectus, ✓ because erectus and sapiens share same genus ✓

Examiner commentary

① Vertebrata is the phylum, not class.

Isla achieves 5/6 marks

Ceri's answer

a) Mammals. ✓

b) A group of organisms that can breed ✗ ① to produce fertile offspring ✓

c) (I) The skulls look similar ✗ ②

(ii) Homo erectus ✓, because the skull looks more human ✗ ③

Examiner commentary

① To be a member of the <u>same</u> species, organisms need to interbreed (or breed together).

② Similar skulls is too vague. Ceri needs to be specific about the shape of the skull, jaw, cranium or teeth, and to include which primates are being referred to.

③ Ceri should have used their classification as asked, rather than referring to the diagram.

Ceri achieves 3/6 marks

Q&A 10

a) The diagram shows a longitudinal section through a part of the alimentary canal.

(i) Name the part of the alimentary canal where structure A would be found. [1]

(ii) Name the blood vessel that transports amino acids to the liver. [1]

(iii) Use the diagram to complete the following table. [3]

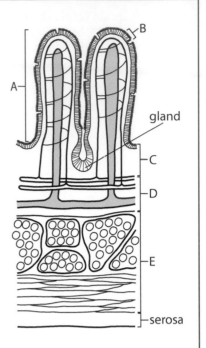

Letter	Name	Function
B		increases surface area
C		contains glands that release secretions
D	submucosa	

b) Coeliac disease is a disease that affects the small intestine, causing the villi to become flattened. Suggest why symptoms often include diarrhoea and fatigue. [3]

Isla's answer

a) (i) Ileum ✓
(ii) Hepatic portal vein ✓
(iii) B = microvilli ✓
C = mucosa ✓
D = contains vessels to transport products of digestion ✓

b) Less glucose can be absorbed for respiration resulting in fatigue. ✓ Diarrhoea results as less water can be absorbed. ✓ ①

Examiner commentary

① To achieve full marks, Isla would need to explain the effect of flattening the villi, i.e. that the surface area is reduced for absorption and digestion (due to membrane bound enzymes).

Isla achieves 7/8 marks

Ceri's answer

a) (i) Ileum ✓
(ii) Hepatic portal vein ✓

b) (iii) B = cilia X ①
C = mucosa ✓
D = contains blood vessels ② X

c) Less glucose can be absorbed resulting in fatigue and diarrhoea ✓ ③

Examiner commentary

① Ceri has confused cilia with microvilli: Cilia are tiny hairs found lining the trachea, whilst microvilli are formed from the in folding of the villi membrane.

② Ceri fails to include the function here, that the vessels transport products of digestion.

③ Ceri needs to make the link between less glucose absorbed and reduced respiration, hence fatigue. Ceri needs to explain why diarrhoea results.

Ceri achieves 4/8 marks

The diagrams below show a section through a healthy lung and a section through a lung from a patient suffering from emphysema. Both diagrams are drawn to the same scale.

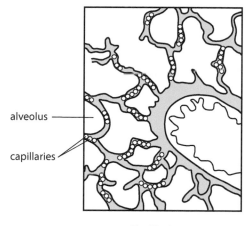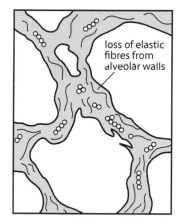

loss of elastic fibres from alveolar walls

alveolus

capillaries

Healthy lung

Emphysema sufferer

a) List **three** adaptations for gas exchange shown in the healthy lung diagram above. [3]

b) Describe how one of the adaptations listed in part (a) aids gas exchange. [1]

c) Using the information in the diagrams, suggest why sufferers of emphysema are often short of breath when performing light exercise. [4]

Isla's answer

a) The walls are thin ✓ and a large network of capillaries can be seen. ✓　①

b) The thin walls reduce the distance that gases have to diffuse. ✓

c) Fewer alveoli reduce the surface area for gas exchange ✓, and because the alveolar walls are thicker ✓, the diffusion distance is increased. ✓ ②

Examiner commentary

① To achieve full marks, Isla would need to include that there is a large surface area.

② Isla needs to make the final link, that between reduced surface area and a reduction in oxygen absorbed. Isla could also have included the effect of a loss of elasticity from the alveolar walls and a reduction in tidal volume/volume of air that could be exchanged in one breath.

Isla achieves 6/8 marks

Ceri's answer

a) The cell walls are thin ✗ ① and moist. ✗ ② a good blood supply can be seen ✗.　②

b) Thin cell walls means that diffusion occurs faster.✗　③

c) Fewer alveoli reduces the surface area for oxygen absorption ✓, and because the walls are thicker ✓, diffusion takes longer. ✗ ④

Examiner commentary

① Ceri has used the term cell wall rather than alveolar walls which is incorrect as animal cells don't have cell walls.

② Ceri has described generic features of gas exchange surfaces, rather than describing those shown in the diagram. Ceri would need to say large capillary network instead.

③ Ceri hasn't been penalised again for the use of cell walls, but would need to say that the diffusion distance is reduced NOT that diffusion occurs faster.

④ Ceri would need to explain that the effect of thicker alveolar walls is that the diffusion distance is increased which would result in less oxygen being absorbed.

Ceri achieves 2/8 marks

Q&A 12

Scientists carried out an experiment to investigate which vessels in a plant were used to translocate solutes. They placed one plant in a gas jar, and added carbon dioxide containing radioactive carbon (^{14}C). After two hours, scientists cut a section of the stem and exposed it to photographic film to produce an autoradiograph.

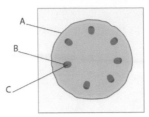

section of stem placed against photographic film in the dark

autoradiograph seen after 12 hours

(a) Identify structures A,B and C. [3]

(b) Using the autoradiograph, and your knowledge, answer the following questions.
  (i) Name the vessel through which solutes are translocated. [1]
  (ii) Explain fully how you arrived at this conclusion. [3]

Isla's answer

a) A= endodermis ✗ ①
  B= Phloem ✓
  C= Xylem ✓
b) (i) phloem ✓
c) (ii) the autoradiograph shows dark areas ✓ in the same place where the phloem is found ✓. ②

Examiner commentary

① Isla has confused endodermis with epidermis

② Isla could have included that the radioactive carbon in carbon dioxide has been converted into sucrose which is then translocated in the phloem vessels

Isla achieves 5/7 marks

Ceri's answer

a) A= epidermis ✓
  B= Xylem ✗
  C= Phloem ✗ ①
b) (i) phloem ✓
c) (ii) there are some dark areas showing that radioactive carbon dioxide must be moving through the vessels ✗ ②

Examiner commentary

① Ceri has confused the location of the xylem with phloem

② Ceri has identified that dark areas are present, but has incorrectly stated that this is due to radioactive carbon dioxide, and has not named which vessel. She should have mentioned that it is radioactive sucrose that is moving through phloem vessels.

Ceri achieves 2/7 marks

Q & A 13

(a) Students were asked to set up an experiment to investigate water loss by a plant. The instructions are given below.
   A. Cut two leafy shoots
   B. Cover the leaves of one shoot with petroleum jelly (Vaseline)
   C. Place the shoots in separate beakers of water and cover surface of water with oil
   D. Record the total mass of each experimental setup
   E. Expose the shoots to light and weigh them again at 30-minute intervals for 5 hours. Then calculate the percentage change in mass.

   The students observed that the percentage change in mass of the shoot with petroleum jelly on its leaves was less than the shoot with no petroleum jelly.

   The students concluded that the percentage change in mass of water lost from the shoot was equal to the mass of water absorbed by the shoot. Explain why the students would be incorrect in reaching this conclusion. [3]

(b) Students also wanted to calculate the average width of an open stomata. Describe how they could do this using a series of electronmicrographs of the underside of a leaf. The magnification on the electronmicrographs was 2500×. [3]

Isla's answer

a) Some of the water taken into the plant would be needed for other things such as photosynthesis ✓ and maintaining turgidity✓.①

b) Measure the width of a stomata on the photograph in mm and convert to μm by multiplying by 1000✓. Then divide this by the magnification, 2500 to find the width of the actual stomata (object) ✓. This would need to be repeated at least 3 times so a mean could be calculated. ✓

Examiner commentary

① Isla could have included that some water is produced in respiration.

Isla achieves 5/6 marks

Ceri's answer

a) Water is also needed for photosynthesis ✓ and maintaining turgidity✓. ①

b) Measure the width of a stomata ② then divide this by 2500 to find the width of the actual stomata ✓ ③

Examiner commentary

① Ceri could also have included that some water is produced in respiration.

② Ceri would need to include details of how to convert to μm.

③ Ceri needs to say how an average (mean) would be calculated by performing repeats.

Ceri achieves 3/5 marks

Q & A 14

Describe how carbon dioxide is carried in the blood. Suggest why oxygen is released more readily in muscles where lactic acid has built up. [9]

Isla's answer

Some carbon dioxide is carried in red blood cells as carbamino-haemoglobin ✓, but most is carried as bicarbonate ions (HCO_3^-) ✓. Carbon dioxide is converted into HCO_3^- inside the red blood cells, following a reaction involving the enzyme carbonic anhydrase ✓. ① HCO_3^- diffuses out of the red cell and into the plasma where it is transported to the lungs ✓. During exercise, muscles begin to respire anaerobically and so glucose is converted to lactic acid which builds up in muscles. ✓ The increased lactic acid concentration in the muscles lowers the pH of blood as protons are released from lactic acid. This reduces haemoglobin's oxygen affinity which is called the Bohr effect ✓, and causes oxyhaemoglobin to dissociate more easily, which is seen by a shift to the right in the oxygen dissociation curve ✓. This is an advantage during exercise, as oxygen is released more readily to respiring tissues ✓. ②

Examiner commentary

Isla constructs an articulate, integrated account, which shows sequential reasoning. The answer fully addresses the question with no irrelevant inclusions or significant omissions. The candidate uses scientific conventions, vocabulary and spelling appropriately and accurately.

① Isla could have included details of the dissociation of carbonic acid to hydrogen and hydrogen carbonate ions and the chloride shift that occurs to maintain electrochemical neutrality via facilitated diffusion.

② Isla could have expanded her answer to include reference to H^+ ions binding to oxyhaemoglobin which releases oxygen.

Isla achieves 7/9 marks

Ceri's answer

Carbon dioxide is produced as a waste gas which is transported in the plasma to the lungs where it is eliminated. Some carbon dioxide is carried in red blood cells as carbamino-haemoglobin. ✓
When exercising, more carbon dioxide is produced which needs to be eliminated. We breathe faster and harder to help to get rid of it. ① Carbon dioxide is carried in the blood dissolved in the plasma. ②. The dissociation curve changes when there is lots of lactic acid, it shifts to the right, something called the Bohr shift. ✓ ③

Examiner commentary

Ceri makes some relevant points, such as those in the indicative content, but shows limited reasoning. The answer addresses the question but with significant omissions. The candidate has limited use of scientific conventions and vocabulary.

① Ceri needs to link high lactic acid concentrations to a fall in pH.

② Ceri should expand how carbon dioxide is carried, i.e. as hydrogen carbonate ions in the blood, and some as carbamino-haemoglobin in the red blood cells.

③ The Bohr shift is mentioned but Ceri needs to demonstrate understanding; for example, the effect on haemoglobin's affinity for oxygen, and how this releases oxygen more easily.

Ceri achieves 3/9 marks

Additional practice questions

1. Inorganic ions are needed by living organisms.

   a) Complete the functions for the following ions in the table below. [2]

Ion	Function
Magnesium	
Phosphate	

   b) Explain why a diet lacking in iron often results in anaemia. [2]

   c) Water is vital to life. Explain how its properties enable it to act as a coolant for animals. [3]

2. a) What is meant by a tissue? [1]

   b) Explain how squamous epithelium is adapted for its function in the lungs. [2]

   c) Penicillin is an antibiotic that works by preventing the cross linking of peptidoglycan in the cell walls of Gram positive bacteria, causing the cells to have weakened cell walls. This makes the bacterium susceptible to osmotic lysis. Using your knowledge of virus structure, explain why antibiotics are ineffective against viruses such as influenza. [1]

3. The diagram shows the fluid mosaic model for membrane structure proposed by Singer and Nicolson in 1972.

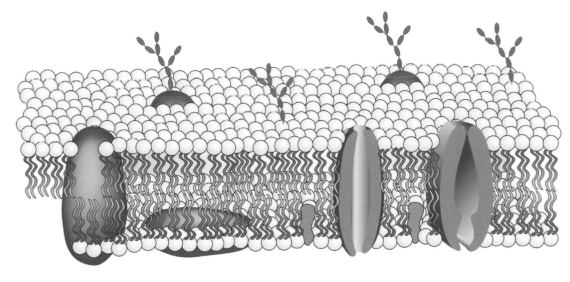

   a) Why is the model said to be fluid mosaic? [2]

   b) In cystic fibrosis, patients have faulty chloride transport proteins in the membranes of the ciliated epithelium that line the trachea and bronchioles. As a result, chloride ions are not secreted from the cell, and are not absorbed by the mucus that covers the cell.

      i) Using your knowledge of osmosis, explain why mucus in these patients is often thick. [2]

      ii) Suggest why patients often report that breathing sea air alleviates their symptoms. [2]

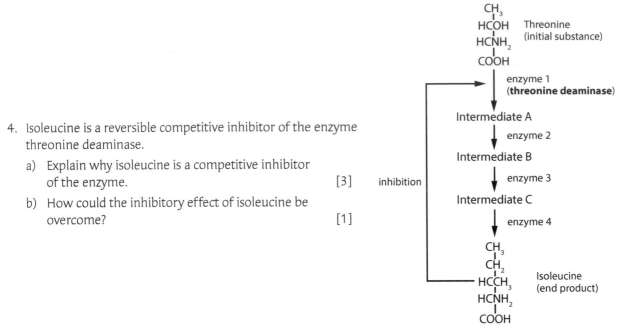

4. Isoleucine is a reversible competitive inhibitor of the enzyme threonine deaminase.

   a) Explain why isoleucine is a competitive inhibitor of the enzyme. [3]

   b) How could the inhibitory effect of isoleucine be overcome? [1]

5. Tetracycline is an antibiotic that binds to the smaller ribosome subunit in bacteria, preventing tRNA from joining the mRNA–ribosome complex. In an experiment, scientists looked at the reduction in growth of bacteria on agar plates where different concentrations of tetracycline were used. The results are shown in the table:

Concentration of tetracycline / mg	Reduction in growth of population / %
0	0
10	15
20	30
40	60

   a) What can you conclude from the results? [2]

   b) How could you refine the experiment to increase the confidence in your conclusion? [2]

   c) Explain why growth of the population is inhibited in bacteria where tetracycline is used. [2]

6. a) The percentage of cells in each stage of the cell cycle is proportional to the length of that stage. Using a microscope a student observed 100 cells and found 5 undergoing prophase. If the total length of the cell cycle is 24 hours, calculate the length of prophase in minutes. Show your working. [2]

   b) Describe how metaphase 1 of meiosis differs from metaphase of mitosis. [2]

7. Describe how you would carry out an experiment to estimate the number of woodlice present in a 100m² woodland. [5]

8. The diagram below shows the tracheal system of an insect.

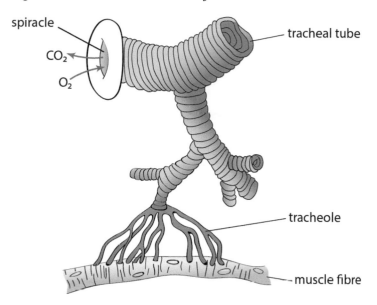

a) Using the diagram and your knowledge, explain why insects are essentially limited in their size and shape. [3]
    b) Explain how insects are adapted for living in dry environments. [2]

9. Kwashiorkor is a condition where fluid is retained, most noticeably in the body of malnourished children, as a result of a very low protein diet. Using your knowledge of osmosis, explain why fluid is retained. [5]

10. The diagram below shows the different pathways by which water crosses the root cortex in plants.

a) Identify pathways A, B and C. [2]
    b) Using your knowledge of the structure of the cell wall, explain why water moves mostly through pathway A. [2]

11. Praziquantel is used to treat tapeworm infections, and is thought to work by the tapeworm losing the ability to resist digestion by the mammalian host. Using your knowledge of *Taenia*, suggest how the drug works to eliminate the parasite. [2]

Quickfire answers

Section 1.1

① Negative charged Cl⁻ ions attract the positive dipole of water, whilst the positive charged sodium ions, Na⁺, attract the negative dipole.

② A polar molecule carries an unequal distribution of electrical charge. The oxygen has a slightly negative charge, whilst the hydrogen is slightly positive.

③ A = cohesion creating surface tension, B = high latent heat of vaporisation, C = solvent

④ A = pentose B = glucose C = triose

⑤ $C_6H_{12}O_6 + C_6H_{12}O_6 - H_2O = C_{12}H_{22}O_{11}$

⑥

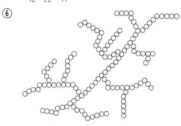

Amylopectin

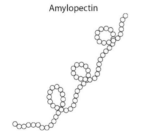

Amylose

⑦ Ester bond, hydrolysis reaction

⑧ Lipids contain twice as much energy as carbohydrates. Lipids don't affect the water potential of cells so are osmotically inert.

⑨ Water produced from the oxidation of food during respiration.

⑩ A phospholipid has 2 fatty acids; triglycerides have 3. A phospholipid has a phosphate group; triglycerides do not.

⑪ Lipid tail of phospholipid

⑫ Peptide bond

⑬ Hydrogen, ionic, disulphide, and peptide

⑭ A and C are globular proteins, B is a fibrous protein.

Section 1.2

① A = 2000, B = 7.25 × 10⁻³, C = 130

② chloroplast

③ 1 = C, 2 = A, 3 = B, 4 = F, 5 = D, 6 = E

④

Prokaryotic cells	Eukaryotic cells
e.g. bacteria and blue-green algae	e.g plants, animals, fungi and protoctists
No membrane-bound organelles	**Membrane-bound organelles**
Ribosomes are smaller (70S) and lie free in cytoplasm	Ribosomes are larger (80S) lie free and attached to membranes, e.g. rough ER
DNA lies free in the cytoplasm	DNA located on chromosomes within the nucleus
No nuclear membrane	**Distinct membrane-bound nucleus**
Cell wall containing peptidoglycan (murein)	Cell wall in plants made of cellulose. In fungi it is made of chitin

⑤ 1 = D, 2 = F, 3 = E, 4 = C, 5 = B, 6 = A

Section 1.3

① Rate of uptake is directly <u>proportional</u> to oxygen concentration.

② Rate of uptake is directly <u>proportional</u> to nitrate ion concentration between points A and B but graph reaches a plateau between B and C as number of carrier proteins becomes limiting.

③ a) This increases the oxygen in the soil, which is needed by root cells to produce ATP during aerobic respiration. ATP is then used for the active transport of mineral ions into the root cells.

 b) Poor growth is the result of the inability of plants to actively uptake nitrate ions due to the lack of oxygen in water logged soils. Plants need nitrates to synthesise proteins and grow.

④ A to C, A to B and C to B
A to B will be quickest – gradient is highest

⑤ As there is no net movement, Ψ = water potential of the soil = −100 kPa.

$-100 = \Psi S + 200$

$\Psi S = -100 - 200 = -300$ kPa

⑥ We can conclude that as the temperature of the beetroot increases, more dye diffuses out. In other words, increasing the temperature increases the permeability of the beetroot membranes. This is because:
Rate of diffusion is increased by increased temperature, because the dye particles have more kinetic energy.
At high temperatures, the proteins in the membrane begin to denature and create gaps, allowing more dye to diffuse out.

Extra

(i) Ethanol dissolves phospholipids which creates holes in the membrane

(ii) Increased temperature increases kinetic energy of the dye molecules which increases rate of diffusion of dye across membrane.

⑦ 1 = D, 2 = A, B, C, D, 3 = C, 4 = B, D (C)

Section 1.4

① Metabolism = anabolism + catabolism

②

Without enzyme

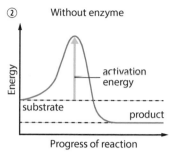

With enzyme

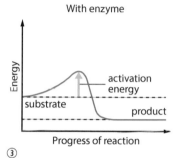

③

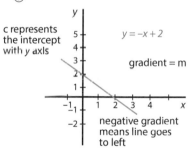

c represents the intercept with y axis

$y = -x + 2$

gradient = m

negative gradient means line goes to left

④ Competitive inhibitors attach to the active site, whereas non-competitive inhibitors attach to an allosteric site (site other than active site).

⑤ (1) Biosensor is specific and will only detect glucose, whereas the Benedict's test could detect the presence of any reducing sugar.

(2) Biosensors can detect far lower concentrations than the Benedict's test.

(3) Result is quantitative, unlike the Benedict's test which is qualitative (relies on a colour change).

Section 1.5

① It must be DNA (due to thymine) and single stranded as A does not equal T.

② CUAAAGGCUUAACCGG (Careful – remember no thymine in RNA)

③ RNA polymerase

④ It contains non-coding regions or introns that must be removed.

⑤ Addition of amino acid to tRNA, which requires ATP.

⑥ 1 = B, 2 = A, 3 = B, 4 = C

Section 1.6

① 1 = B, 2 = A, 3 = D, 4 = C, 5 = C

② Correct order should be D, C, B, A

③ Any three from DNA replication, growth, organelles synthesised, ATP and protein synthesis.

④ 1 = A, 2 = A, 3 = B, 4 = C, 5 = C and E

⑤ A = anaphase 1, B = metaphase 1, C = anaphase 2, D = interphase (allow early prophase)

Section 2.1

① Homologous structures, e.g. forelimb of whale and human have a similar structure and therefore origin, whereas analogous structures have a similar function but do not share the same origin, e.g. wings of a bird and insect.

② Field A has the greater biodiversity as it is closer to 1.

③ A. False
B. True
C. True
D. True

Section 2.2

① Increasing size increases metabolic rate, so the oxygen requirements increase, but the surface area : volume ratio is decreasing, so the external surface is insufficient to meet the requirements.

② Gills, lungs, tracheae

③ less, same, parallel, part, equilibrium, less, ventilate, floor, swimming

④ pulmonary artery

⑤ A = FALSE, B = FALSE, C = TRUE, D = FALSE, E = TRUE

⑥ 15/50 = 0.3mm

Extra
5.1 kPa. Equilibrium reached, i.e. partial pressures the same.

Section 2.3a

① A. Capillary
B. Artery
C. Artery, capillary, vein
D. Artery, vein
E. Vein

② P = Voltage change associated with contraction of atria
QRS = Contraction of ventricles
T = Repolarisation of ventricle muscles
TP = The filling time

③ Shift to left = haemoglobin has a *higher* affinity for oxygen and so becomes more saturated with oxygen than normal at the same low partial pressure of oxygen. Shift to right = haemoglobin has a *lower* affinity for oxygen and so releases its oxygen more easily than normal at the same low partial pressure of oxygen.

④ More haemoglobin/red blood cells produced.

Extra
95 – 44% released = 51%
$0.51 \times 280 \times 10^6 = 142.8 \times 10^6$
$142.8 \times 10^6 \times 4 = 571.2 \times 10^6$

⑤ Maintains electrochemical neutrality.

Section 2.3b

① Support, and transport of water (and mineral ions).

② The Casparian strip forms a waterproof band around the endodermal cells, which prevents water passage.

③ It would be reduced or prevented as mineral uptake requires ATP for active transport.

④ Adhesion, cohesion, capillarity, root pressure.

⑤ A. Decreases – saturated air is not blown away from leaf surface/ diffusion shells remain which decreases water potential gradient.

   B. Decreases – rain increases humidity so decreases gradient.

C. Increases – increased kinetic energy increases rate of evaporation and diffusion.

⑥ Air surrounding stomata is saturated with water vapour reducing water potential gradient between inside and outside of leaf.

⑦ Both have waterproof cuticle and can close spiracles/stomata to reduce water loss.

⑧ Produce ATP for active transport.

⑨ Any two from: sieve tubes, companion cells, phloem parenchyma.

⑩ Growing points, e.g. roots.

Section 2.4

① Mucosa, submucosa, muscle and serosa.

② A and C = Capillary, B = Lacteal

③ Herbivore – canines indistinguishable from incisors, large molars with interlocking enamel ridges.
   Carnivore – has carnassial teeth, and large canines.

④ Mammals do not produce cellulase. Without the bacteria, they could not digest cellulose.

⑤ Extremes of pH, digestive enzymes, host's immune system, peristalsis, difficulty in finding a mate.

See answers to digestion summary on page152.

Answers – additional practice questions

Mark scheme

1. a) Magnesium – constituent of chlorophyll [1]
        Phosphate – a constituent of phospholipids in cell membranes / needed to make nucleotides/ATP [1]

    b) Iron is needed to manufacture haemoglobin [1]

        Without it, fewer red blood cells can be made [1]

    c) Water has a high latent heat of vaporisation due to the large number of hydrogen bonds between molecules [1]
        When water evaporates from sweat a large amount of energy is needed to make water evaporate / break hydrogen bonds between water molecules [1]
        Ref to water being a dipolar molecule [1]

2. a) A group of similar cells working together to perform a particular function [1]

    b) Consists of flattened cells [1]
        creates a short diffusion pathway for diffusion of gases/or named gas [1]

    c) Viruses do not possess peptidoglycan/cell walls so antibiotics have no effect [1]

3. a) Fluid – phospholipids are free to move [1]

        Mosaic – protein molecules are randomly assorted [1]

    b) i)   Chloride ions don't lower water potential of mucus / mucus water potential remains high [1]

             Water does not enter mucus by osmosis/down water potential gradient so remains thick [1]

         ii) Breathe in sea air / sodium chloride [1] which lowers water potential of mucus so water enters it by osmosis, thinning it [1]

4. a) Isoleucine has a similar shape to threonine [1] so competes for the enzyme's active site [1] preventing the substrate/threonine from binding [1]

   b) When concentration of isoleucine falls/threonine concentration increases [1]

5. a) Tetracycline reduces bacterial growth [1], doubling the concentration of tetracycline, doubles the reduction in growth [1]

   b) More concentrations of tetracycline [1] use of repeats [1] NOT use a control

   c) Translation of mRNA is prevented by tetracycline [1] so proteins cannot be synthesised for (growth / example of protein) [1]

6. a) 5/100 [1] × (24 × 60) = 72 minutes [1]

   b) Metaphase 1 (meiosis) homologous pairs of chromosomes (bivalents) align at the equator, whereas in metaphase of mitosis, chromosomes align [1]

      Metaphase 1 (meiosis) involves independent assortment of chromosomes, which does not occur in metaphase of mitosis [1]

7. Animals are captured from a set area, e.g. 10m², and marked (it is important that they are not harmed or made more visible to predators) and then released. [1] Once they have had chance to reintegrate with the population, e.g. 24hrs, they are recaptured. [1] The total population size can be estimated using the number of individuals captured in sample 2, and the number in that sample that are marked (i.e. caught before). [1]

   Pop size = no. in sample 1 × no. in sample 2 / no. marked in sample 2 [1]

   Reference to repeating the experiment [1]

   Need to multiply up from 10m² to 100m² [1] ANY 4

   Have to assume that no births/deaths/immigration/emigration, have occurred during the time between collecting both samples [1]

8. a) Gas exchange system is limited, e.g. no respiratory pigment / blood for transport of oxygen/carbon dioxide [1]

      Rely on diffusion of gases through tracheal system, which is slow [1]

      Ventilation mechanism relies on abdominal movements [1]

      gases have to dissolve into fluid at end of tracheoles before diffusing into muscle [1]

      ANY 3

   b) Can close spiracles to reduce water loss [1]

      Exoskeleton contains chitin, which is waterproof/ ref to wax layer on exoskeleton [1]

9. Low protein diet results in decrease blood albumin/blood protein [1] which raises the water potential of the blood [1]. This reduces the water potential gradient at the venous end of the capillary bed/network [1] so less water is reabsorbed into the blood by osmosis [1]. There is more excess tissue fluid which cannot drain into lymphatic system so fluid accumulates in body tissues [1]

10. a) A = apoplast, B = symplast, C = vacuolar pathways. (3 correct = 2, 2 correct = 1 mark)

    b) (Apoplast/pathway A) water moves between spaces in the cellulose cell wall [1], this is possible due to cellulose fibres being freely permeable to water and solutes [1]

11. Prevent synthesis of enzyme inhibitors [1], thins cuticle [1]

Digestion summary answers.

Food	Region of gut	Enzyme(s)	Site of production	pH	Substrate	Products	How Absorbed
Carbohydrate	Mouth	Amylase	**Salivary glands**	7	**Starch**	Maltose	
	Duodenum (1st part of small intestine)	Amylase	**Pancreas**	7	**Starch**	Maltose	
	Ileum (2nd part of small intestine)	**Maltase**	**Ileum** mucosa	8.5	Maltose	**Glucose**	Glucose enters by **co-transport** into **epithelial** cells and by **facilitated diffusion** into capillary of villus
		Sucrase			Sucrose	**Glucose + fructose**	
		Lactase			Lactose	**Glucose + galactose**	
Protein	Stomach	**Peptidase**	Gastric glands	2	**Protein**	Polypeptides	
	Duodenum (1st part of small intestine)	endopeptidases	**Pancreas**	7	**Protein**	Polypeptides	
	Ileum (2nd part of small intestine)	endopeptidases & exopeptidases	**Ileum** mucosa	8.5	Polypeptides	**Amino acids**	Amino acids enter by active transport into **epithelial** cells and then by **facilitated diffusion** into the capillary of villus

Food	Location	Enzyme	Source	pH	Substrate	Product	Absorption / Notes
Lipid*	Duodenum (1st part of small intestine)	Lipase	Pancreas	7	Lipids	Fatty acids & glycerol	Fatty acids and glycerol enter **epithelial** cells via diffusion. They recombine into **triglycerides** and enter lacteal of villus
	Ileum (2nd part of small intestine)	Lipase	Ileum mucosa	8.5	Lipids	Fatty acids & glycerol	
Cellulose	-NA-	-	-	-	-	-	Provides bulk and stimulates **peristalsis**
Water	-NA-	-	-	-	-	-	By **osmosis** into villi and colon

Index